T0212343

Politics and Policies of Rural Authenticity

This book explores the notion of rurality and how it is used and produced in various contexts, including within populist politics which derives its legitimacy from the rural–urban divide.

The gap between the "common people" and the "elites" is widening again as images of rurality are promoted as morally pure, unalienated and opposed to cultural and economic globalisation. This book examines how using certain images and projections of rurality produces "rural authenticity", a concept propagated by various groups, including regional food producers, filmmakers, policymakers and lobbyists. It seeks to answer questions such as: What is the rurality that these groups of people refer to? How is it produced? What are the purposes that it serves? The research in this book addresses these questions from the areas of both politics and policies of the "authentic rural". The "politics" refers to polarisations including politicians, social movements and political events which accentuate the rural–urban divide and bring it back to the core of the societal conflict, while the "policies" focus on rural tourism, heritage industry, popular art and other areas where rurality is constantly produced and consumed.

With international case studies from leading scholars in the field of rural studies, the book will appeal to geographers, sociologists and politicians, as well as those interested in the re-emergence of the rural–urban divide in politics and media.

Pavel Pospěch is Associate Professor of Sociology at Masaryk University in Brno, Czech Republic, and a Faculty Fellow of the Yale Center for Cultural Sociology. He has published on topics in urban sociology, rural sociology and cultural theory.

Eirik Magnus Fuglestad is a researcher at Ruralis – Institute for Rural and Regional Research, Trondheim, Norway. His research interests lie in the intersection between nationalism, property rights and rural and agricultural development, looking at these fields from a historical sociological perspective.

Elisabete Figueiredo is Associate Professor with Habilitation at the Department of Social, Political and Territorial Sciences and Full Researcher at GOVCOPP (Research Unit in Governance, Competitiveness and Public Policies), University of Aveiro, Portugal. She has published in topics including rural sociology, environmental sociology and risk perceptions.

Perspectives on Rural Policy and Planning

This well-established series offers a forum for the discussion and debate of the often-conflicting needs of rural communities and how best they might be served. Offering a range of high-quality research monographs and edited volumes, the titles in the series explore topics directly related to planning strategy and the implementation of policy in the countryside. Global in scope, contributions include theoretical treatments as well as empirical studies from around the world and tackle issues such as rural development, agriculture, governance, age and gender.

The Changing World of Farming in Brexit UK
Edited by Matt Lobley, Michael Winter and Rebecca Wheeler

Rural Gerontology
Towards Critical Perspectives on Rural Ageing
Edited by Mark Skinner, Rachel Winterton and Kieran Walsh

Tourism and Socio-Economic Transformation of Rural Area
Evidence from Poland
Edited by Joanna Kosmaczewska and Walenty Poczta

Governance for Mediterranean Silvo-Pastoral Systems
Lessons from the Iberian Dehesas and Montados
Edited by Teresa Pinto-Correia, Helena Guimarães, Gerardo Moreno and Rufino Acosta Naranjo

Politics and Policies of Rural Authenticity
Edited by Pavel Pospěch, Eirik Magnus Fuglestad and Elisabete Figueiredo

For more information about this series, please visit: www.routledge.com/Perspectives-on-Rural-Policy-and-Planning/book-series/ASHSER-1035

Politics and Policies of Rural Authenticity

Edited by
Pavel Pospěch, Eirik Magnus Fuglestad
and Elisabete Figueiredo

Routledge
Taylor & Francis Group
LONDON AND NEW YORK

First published 2022
by Routledge
2 Park Square, Milton Park, Abingdon, Oxon OX14 4RN

and by Routledge
605 Third Avenue, New York, NY 10158

*Routledge is an imprint of the Taylor & Francis Group, an informa
business*

© 2022 selection and editorial matter, Pavel Pospěch, Eirik Magnus
Fuglestad, Elisabete Figueiredo; individual chapters, the contributors

The right of Pavel Pospěch, Eirik Magnus Fuglestad, and Elisabete
Figueiredo to be identified as the authors of the editorial material,
and of the authors for their individual chapters, has been asserted in
accordance with sections 77 and 78 of the Copyright, Designs and
Patents Act 1988.

British Library Cataloguing-in-Publication Data
A catalogue record for this book is available from the British Library

Library of Congress Cataloging-in-Publication Data
A catalog record has been requested for this book

ISBN: 978-0-367-55044-8 (hbk)
ISBN: 978-0-367-55045-5 (pbk)
ISBN: 978-1-003-09171-4 (ebk)

DOI: 10.4324/9781003091714

Typeset in Times New Roman
by SPi Technologies India Pvt Ltd (Straive)

Contents

List of Illustrations   vii
List of Contributors   viii

1   **Rural authenticity between commodification and populism**   1
EIRIK MAGNUS FUGLESTAD, ELISABETE FIGUEIREDO AND PAVEL POSPĚCH

PART I
Politics of rural authenticity   13

2   **City and countryside in the imagining of nations**   15
JONATHAN HEARN

3   **Revisiting the politics of the rural and the Brexit vote**   27
MICHAEL WOODS

4   **Populism of the dispossessed: Rethinking the link between rural authenticity and populism in the context of neoliberal regional governance**   42
BIANKA PLÜSCHKE-ALTOF AND AET ANNIST

5   **The minister's tears and the strike of the invisible: The political debate on the "regularisation" of undocumented migrant farm labourers during the Covid-19 health crisis in Italy**   60
DOMENICO PERROTTA

6   **Political and apolitical dimensions of Russian rural development: Populism "from above" and *narodnik* small deeds "from below"**   77
ALEXANDER NIKULIN AND IRINA TROTSUK

7   The feeling of being robbed   94
BJØRN EGIL FLØ

PART II
Policies of rural authenticity   111

8   #Proudofthefarmer: Authenticity, populism and rural
    masculinity in the 2019 Dutch farmers' protests   113
    ANKE BOSMA AND ESTHER PEEREN

9   Idyllic politics and politics of the idyll   129
    FLORIAN DÜNCKMANN

10 Dystopia as authenticity: Changing ruralities in Icelandic cinema 141
    THORODDUR BJARNASON, BRYNHILDUR THORARINSDOTTIR AND
    MENELAOS GKARTZIOS

11 Rural authenticity as cosmopolitan modernity? Local political
    narratives on immigration and integration in rural Norway   155
    GURO KORSNES KRISTENSEN AND BERIT GULLIKSTAD

12 Dynamics of changes in the farmers' contestation in Poland in
    1989–2018: On the way to rationality and an institutionalised
    model of collaboration   170
    GRZEGORZ FORYŚ

13 Rurality: From the margins to the focus of interest   187
    PAVEL POSPĚCH, EIRIK MAGNUS FUGLESTAD AND
    ELISABETE FIGUEIREDO

    Index   195

Illustrations

Figure

10.1 The production of urban and rural films in Iceland and
      proportion of the population living in the Reykjavík
      Capital Area, 1918–2019   146

Tables

3.1 Perceptions of members of the Countryside Alliance on
     rural–urban relations, 2007   35
3.2 Perceptions of members of the Countryside Alliance on
     political institutions, representation and radicalism, 2007 38
4.1 List of cited interview partners   46
12.1 Opinions on the existence of organisations defending
      farmers' interests and the most effective forms of defence
      of farmers' interests (in %)   179

Contributors

Aet Annist is a social anthropologist, Associate Professor at the University of Tartu and Senior Researcher at the Tallinn University in Estonia. She has published articles and books on various aspects of post-socialist transformations, migration, institutionalisation of heritage and more recently on protest movements, linking these themes to changes in social mutuality. Her most recent focus is on socialities arising from climate related concerns.

Thoroddur Bjarnason is Professor of Sociology at the University of Akureyri, Iceland. He holds an MA from the University of Essex and a PhD from the University of Notre Dame. His research interests include patterns of inclusion and exclusion, urban–rural relations, and causes and consequences of geographical mobility. Recent publications include contributions to Culture, Health and Sexuality, Ethnic and Racial Studies, Journal of Rural Studies, International Journal of Circumpolar Health, Population, Space and Place, and Sociologia Ruralis.

Anke Bosma is a PhD candidate in the ERC-funded project "Imagining the Rural in a Globalizing World" (2018–2023) at the Amsterdam School for Cultural Analysis (ASCA) at the University of Amsterdam. She explores how the rural is imagined in the Dutch cultural consciousness in relation to globalisation. Another area of interest is (de)colonial theory and history in the Netherlands, which she developed while doing research at the Research Centre for Material Culture.

Florian Dünckmann is a Professor of Cultural Geography at Kiel University. His fields of research are Politics of Rurality, Practice Theory and Political Ecology. Currently, he is developing a Geography of Political Practices based on the work of Hannah Arendt. He has conducted several research projects, investigating, for example, micro-politics in small rural communes, the role of civic engagement in rural development and the geography of remembering the Cold War.

Elisabete Figueiredo is Associate Professor with Habilitation at the Department of Social, Political and Territorial Sciences and Full Researcher at GOVCOPP (Research Unit in Governance, Competitiveness and Public Policies) University of Aveiro, Portugal. She has published in topics including rural sociology, environmental sociology and risk perceptions

Bjørn Egil Flø holds a PhD in Rural sociology; he works as Senior Researcher at the Norwegian Institute of Bioeconomy Research (NIBIO). For more than 20 years he has studied numerous rural communities all over Norway, mostly conducting ethnographic studies. More recently, he has established himself as one of the leading voices in the Norwegian debate on rural issues.

Grzegorz Foryś is a sociologist and a professor at Pedagogical University of Cracow (Institute of Political Science and Administration). His scientific interests concentrate on social changes, social movements and rural sociology. He has recently been investigating the role of different social movements in promoting multi-functional rural development in Europe, and especially in Poland, from the perspective of sociology and political science.

Eirik Magnus Fuglestad is a researcher at Ruralis – Institute for Rural and Regional Research, Trondheim, Norway. His research interests lie in the intersection between nationalism, property rights and rural and agricultural development, looking at these fields from a historical sociological perspective.

Menelaos Gkartzios is Reader in Planning & Rural Development at Newcastle University's Centre for Rural Economy in the UK. His research has focused on mobilities and social change, housing, the relationship between art and development, and international comparative research. Menelaos has been appointed Visiting Associate Professor at the University of Tokyo in Japan and Visiting Research Fellow at the National University of Ireland, Galway, and at the Univeristy of Akureyri, Iceland. He currently serves as Associate Editor in Habitat International.

Berit Gullikstad is Professor of Gender Studies, Department of Interdisciplinary studies of Culture, Norwegian University of Science and Technology (NTNU), Norway. She has specialised in research on historical and contemporary working life and in the context of the welfare state, with a special focus on gender/ethnicity and intersectional approaches. Gullikstad has managed several research projects, among them Living Integration: At the Crossroad between Official Policies, Public Discourse and Everyday Practices.

Jonathan Hearn is Professor of Political and Historical Sociology at the University of Edinburgh, and President of the Association for the Study of Ethnicity and Nationalism. The author of four books, his work focuses on nationalism, power and the evolution of liberal forms of society, combining theoretical, historical and ethnographic research. He is currently working on a book about the institutionalisation of competition in modern society.

Guro Korsnes Kristensen is Professor in Gender Equality and Diversity Studies at the Department of Interdisciplinary Studies of Culture, the Norwegian University of Science and Technology (NTNU), Norway. Her research focuses on migration, gender, qualitative methods and biopolitics, and she has published several journal articles and book chapters in these fields.

Alexander Nikulin is head of the Center for Agrarian Studies, Russian Presidential Academy of National Economy and Public Administration, and head of the Chayanov Research Centre, Moscow School of Social and Economic Sciences, in Moscow, Russia. He is an author and co-author of books and articles on rural sociology and the agrarian history of Russia. He is Editor-in-Chief of the journal *Russian Peasant Studies*. His recent books include *Agrarians, Power and Countryside: From Past to Present* (2014) and *Chayanov's School: Utopia and Rural Development* (2020).

Esther Peeren is Professor of Cultural Analysis at the University of Amsterdam and Academic Director of the Amsterdam School for Cultural Analysis (ASCA). She directs the ERC-funded project "Imagining the Rural in a Globalizing World" (2018–2023). Recent publications include the edited volume *Other Globes: Past and Peripheral Imaginations of Globalization* (Palgrave, 2019, with Simon Ferdinand and Irene Villaescusa-Illán) and articles in *Landscape Research* and the *Journal of Rural Studies*.

Domenico Perrotta is Assistant Professor in Sociology of Cultural Processes at the University of Bergamo, Italy. His research interests include migrant labour and the transformations of agri-food systems. Among his recent publications include the collective volume *Migration and Agriculture: Mobility and Change in the Mediterranean Area* (Routledge, 2016), co-edited with A. Corrado and C. De Castro.

Bianka Plüschke-Altof is a Researcher in Environmental Sociology and Human Geography at the School of Natural Sciences and Health at Tallinn University, Estonia. A dominant theme in her research relates to questions of socio-spatial and environmental justice in Central-Eastern Europe. Her chapter in this volume is partly based on her doctoral research on peripheralisation and territorial stigmatisation in rural Estonia, defended in 2017 at the University of Tartu. Her most recent focus lies on spatial injustices arising from human–nature relations in the city.

Pavel Pospěch is Associate Professor of Sociology at the Masaryk University in Brno, Czech Republic, and a Faculty Fellow of the Yale Center for Cultural Sociology. He has published on topics in urban sociology, rural sociology and cultural theory. He is Editor-in-Chief of Sociální studia/Social Studies.

Brynhildur Thorarinsdottir is Associate Professor of Icelandic Literature at the University of Akureyri, Iceland. She is also a prolific writer and award-winning author of books for children and teens. She holds a MA from the University of Iceland and a diploma in teaching from the University of Akureyri. Her research interests are in language, culture and education, including medieval literature, adaptations, reading habits and Icelandic language in primary and secondary schools.

Irina Trotsuk is Professor of Sociology, RUDN University; and Senior Researcher of the Center for Agrarian Studies, Russian Presidential Academy of National Economy and Public Administration, of the Chayanov Research Center, Moscow School of Social and Economic Sciences, and of the Center for Fundamental Sociology, Higher School of Economics, in Moscow, Russia. She is the author and co-author of books and articles on sociological methodology and rural sociology. Her recent works include translations into Russian of books by H. Bernstein (2016), J.D. van der Ploeg (2017), and J.C. Scott (2017; 2020).

Michael Woods is Professor of Human Geography and Co-Director of the Centre for Welsh Politics and Society at Aberystwyth University in the UK. He has researched rural politics in Britain and other countries for more than two decades, and is author *of Contesting Rurality: Politics in the British Countryside* (Ashgate, 2005) and editor of *New Labour's Countryside* (Policy Press, 2008), His current research includes examining connections between rural areas, globalisation and the rise of populism.

1 Rural authenticity between commodification and populism

Eirik Magnus Fuglestad, Elisabete Figueiredo and Pavel Pospěch

The idea for this book came about during the XXVIII European Society for Rural Sociology Congress "Rural Futures in a Complex World", held in Trondheim, Norway, 25–28 June 2019. The editors organised a working group at the congress titled "Politics and policies of rural authenticity and the return of nationalism and populism" in which a range of interesting themes and topics were presented and discussed. The topics discussed in the working group were then, and still are, at the forefront of political and socio-economic developments in Europe and the United States, where national populism and issues related to rural–urban divides and centre–periphery relations converge and shape our societies more powerfully than before. For this reason, the editors felt it was timely to collect those discussions and findings in a volume. However, for a number of reasons, not all the presentations from the working group at the European Rural Sociology Congress are included in this book, and the book includes some chapters that were not originally presented at the conference.

The chapters included in this book focus on how the notions of rurality and rural authenticity are used and produced in various discursive framings, including prominently the current wave of nationalist-populist politics, which seeks to build its legitimacy on the long-lasting rural–urban dichotomy. The populist-induced split between "common people" and "elites" attaches itself to the rural–urban divide, and its supporters promote various images of rurality. Often, the rural is portrayed as morally pure, unalienated, upholding national heritage and soul, and opposed to the cultural and economic glo-balisation. There are many different producers of "rural authenticity": the concept is employed and filled with meanings by groups from regional food producers, through film-makers, to policymakers and lobbyists. What is the rurality that these groups refer to? How is it produced, and what purposes does it serve? In this volume, we provide answers from the areas of both the politics and the policy of the "authentic rural".

Discussions on whether rural areas are somehow distinct and meaningfully different from their urban counterparts have been the subject of a long

DOI: 10.4324/9781003091714-1

controversy in rural and urban sociology alike (Sorokin and Zimmerman 1929; Wirth 1938; Simmel 1950; Bell 1992). Generally speaking, the concept of rural distinctiveness has come to be seen as increasingly less plausible: this questioning was spearheaded by the empirical critiques of the notion of rural–urban continuum and by rural sociologists working in the political economy tradition (Pahl 1966; Buttel and Newby 1980; Friedland 1982). Yet in recent years we have witnessed a notable return of rural distinctiveness. Firstly, by a re-emergence of the rural–urban dichotomy through the growing hegemonic social and political representations of the rural as a mainly "authentic" and "idyllic" place (Halfacree 1993; Bell 2006; Soares da Silva et al. 2016), even if these representations do not match the reality of many rural territories (Cloke 2006; Halfacree 2007; Figueiredo 2013). We argue that this resurgence of the rural as a distinct place has also found its way into policy and in politics. "Authentic rurality" became a buzz-term around which public policies, market strategies and cultural productions gravitate. The growth of rural tourism and heritage industries, together with amenity migration and counter-urbanisation, the branding and marketing of local crafts and products and the marketisation of rural identities, as well as cultural products (films, TV shows and websites), produce and reproduce images of "authentic rurality" and turn rural territories into "consuming idylls" (Halfacree 2006: 57). All these phenomena co-produce a matrix of meanings which emphasise the distinctiveness and the value of authentic rurality. In this introduction, we will give a short overview of the major political and economic trends that frame this development. We also provide a brief overview and discussion of the chapters presented in this book.

Rural authenticity in a context of rising nationalist and populist sentiments

Rural authenticity has taken new political forms in a trend that can be observed in recent years. While a politics of the rural, from agrarian parties to rural protest, has been a stable feature of the western political environment, there is now a new sense of urgency to these issues. In the Anglo-American world especially, the Brexit referendum in the UK and the election of Donald Trump as US president in 2016 represented dramatic political shifts where the rural–urban dichotomy played a significant role. The relative success of Marine Le Pen in the French presidential election the following year and the emergence of the *Gilets Jaunes* (yellow vests) movement in France in 2018 further reinforced the impression that the rural was back with new force.

A few key books were quickly established as a sort of canon for an initial understanding of these changes to western European politics. In 2016 Christophe Guilluy published his book with the telling title: *Twilight of the Elites: Prosperity, the Periphery and the Future of France* (English translation published 2019). One of the central points of this book was that new tensions between the large cities – what Guilluy termed "the new citadels" –

and the peripheries were rising in France in a context wherein peripheries and rural areas were increasingly marginalised.[1] In the aftermath of Brexit the following year, the English writer David Goodhart (2017) published *The Road to Somewhere: The Populist Revolt and the Future of Politics*, in which he made the distinction between "somewheres" and "anywheres". This became a tool for understanding the Brexit vote, where the more cosmopolitan, mobile and highly educated upper-middle-class "anywheres" represented the remainers, and the more rooted (often rural or from the peripheries) and less educated "somewheres" represented the leavers. In America, even before the election of Donald Trump as president, Kathrine Cramer published *The Politics of Resentment: Rural Consciousness in Wisconsin and the Rise of Scott Walker* (2016). Cramer showed, among other things, how many rural people attributed rural deprivation to decisions made by urban elites, thus reinforcing the urban–rural dichotomy and creating a seedbed for populist politics.

The trends observed in the aforementioned books are not unique to these three cases: they can be observed throughout Europe, to varying degrees, and they are part of a larger canvas that displays a transformation of politics across Europe and the Americas. This transformation can be interpreted in different ways, but two prisms have been dominant: this is seen either as a return to nationalism, or as a revival of populist politics – or sometimes both. Academics have asked questions such as "why has nationalism not run its course?" (Harris 2016), or "why has nationalism been revived in Europe?" (Hosking 2016). In 2019, a group of intellectuals and academics published a manifesto in major European newspapers pointing to what they saw as the populist nature of this nationalism: "let's reconnect [...] with our 'national soul'! Let's rediscover our 'lost identity'! This is the agenda shared by the populist forces washing over the continent" (The Guardian 2019). There is an ongoing discussion as to whether nationalism and populism can be conflated in the present political climate, or if nationalism and populism should be kept as distinct concepts in this analysis. In the wake of this discussion, Rogers Brubaker has suggested that populism and nationalism should be seen as "analytically distinct but not analytically independent: as intersecting and mutually implicated though not fully overlapping fields of phenomena" (Brubaker 2020: 45). Brubaker further suggest that what is now happening is that "Nationalist discourse claims to restore 'ownership' of the polity to the nation, just as populist discourse claims to restore power to the people" (Brubaker 2020: 51). These two impulses have strengthened and to some degree converged over much of Europe and America during the last decade. The politics of rural authenticity has been part of this, as the chapters of this book will demonstrate. Each chapter in this volume provides insights into different cases related to this development, and the individual chapters use their own analytical frames. In the following, we will provide a broad analytical context within which we may understand the new politics of rural authenticity and the way in which such politics have become entangled with nationalism and populism.

Rising inequalities: the transformation of nations and centre–periphery relations

The trend we are witnessing in European and American politics today, whereby rural authenticity is being revitalised in the context of nationalism and populism, can be seen as a new stage in the development of the liberal nation-state. To frame this, we first invoke two classics in the literature on the emergence of the nation-state. The first one is Ernest Gellner's influential theory of nationalism put forth in its earliest form in *Thought and Change* (1964) and cemented with the publication of *Nations and Nationalism* (1983). For Gellner, the key term when it comes to the creation of nation-states is "homogenisation". Nation-states, according to Gellner, create cultural homogeneity that fosters social mobility and economic growth, and they come into being due to the conditions required by industrial society (Gellner 2006). This symbiosis of nationalism and industrialism created societies that for the first time in history sustained perpetual economic growth and mass social mobility. Gellner thus implied, in the words of John Hall, that a "modern social contract existed in which legitimacy would be and should be given to social orders that provided wealth and in which rule was exercised by those co-cultural with the majority of the population" (Hall 2019: 46). This modern social contract entailed an expectation that the economic growth created should benefit all members of the nation – not necessarily equally, but to at least some extent.

Gellner did not pay much attention to the processes by which such a distribution came into being as nation-states formed. For this we go to the other classic author on the field: the political sociologist Stein Rokkan. Different from Gellner's almost ahistorical models for nation building, Rokkan paid a great deal of attention to the specific historical developments of nation-states, most notably in his essays "Nation building, cleavage formation and the structuring of mass politics" (1970) and "Dimensions of state formation and nation building: a possible paradigm for research on variations within Europe" (1975). Rokkan was in line with Gellner on the homogenising effects of the nation-state, but he emphasised much more strongly the conflict dimensions that this process both created and sustained within nation-states, particularly between the nation's cultural and geographic centre and its peripheries. Nations seek homogeneity, but they are not homogenous: they contain within themselves a variety of subcultures and interest in the peripheries which come to expression through the way party politics is organised. For national homogenisation to be successful, the peripheries not only had to be included in a national culture, they also had to partake in the economic growth and the social mobility that the nation-state made possible (Rokkan 1987). The key point for us here is that Rokkan pointed out how crossing conflict lines that included both class and centre–periphery relations had played out historically to form liberal democracies and redistributive mechanisms within the western nation-states. And, most importantly, while Gellner's modern social contract on perpetual growth and social mobility kept such conflicts more or less contained until recently, we are now seeing a crack in this contract. Centre–periphery relations are being reactivated, and the politics of rural authenticity plays a key role in this.

What has happened, in the words of John Hall, is that "it is no longer the case that one's piece of the pie, however small, will increase year by year" (Hall 2019: 52). The two recent large volumes by Thomas Piketty, *Capital in the Twenty first Century* (2014) and *Capital and Ideology* (2020), have provided rich empirical data on the increase in inequality all over the world. While Europe and America saw unprecedented economic growth and a high degree of economic equality and social mobility in the years following the Second World War, we have witnessed a massive rise in inequality since the 1980s, as Piketty clearly demonstrates. Piketty attributes this to what he calls "hypercapitalism" or "neo proprietarianism". It is important to note that this current hypercapitalism is a global phenomenon, and that it works in an increasingly connected global economy. Piketty points to a change in the political landscape that has occurred alongside the economic and ideological shift towards globalised hypercapitalism: the traditional left and the traditional right political parties have changed in structure and ideology into what he calls the "Brahmin left" (mostly constituted by the higher-educated elites) and the "merchant right" (for the elite rich). Thus, the large groups of working- and middle-class voters are left without political parties that represent them; instead, the Brahmin left and the merchant right display many converging interests in terms of the new globalised political and economic changes (Piketty 2020). This has activated centre–periphery relations (Rokkan's old conflict lines). In this context of a globalised capitalist economy, many rural areas or peripheral areas have become what Andrés Rodríguez-Pose calls "places that don't matter". Rodríguez-Pose writes that this is about:

> how the long-term decline of formerly prosperous places, disadvantaged by processes that have rendered them exposed and almost expendable, has triggered frustration and anger. In turn, voters in these so-called "places that don't matter" have sought their revenge at the ballot box.
>
> (Rodríguez-Pose 2020: 7)

This has resulted in conflicts "with strong territorial, rather than social foundations" (Rodríguez-Pose 2018: 189). The political scientist Francis Fukuyama has further noted how such place-based revolts from the peripheries and from rural areas have contributed to the emergence of politics of identity and authenticity and a reconfiguration of politics in liberal democracy in general (Fukuyama 2018). We can say that what is happening on a greater scale is that one type of nation-state – the liberal nation-state that has dominated Europe and America since Francis Fukuyama's infamous "end of history" moment in 1991 (Fukuyama 1992), which had as its central ideological foundations the Habermasian ideas of civic, constitutional patriotism and global free movement of capital and people – is under pressure from the emergence of new forms of nationalism and populism (Eatwell and Goodwin 2018; Mounk 2018; Hall 2019).

The politics of rural authenticity play a central part in this. From the Trump administration in the USA through Le Pen in France, Orbán in Hungary and Kaczynski in Poland, political leaders have been conjuring images of the authentic rural folk, frequently contrasted with the urban elites, and paintin themselves as champions of the rural people. This reinvention of the rural–urban divide often presents a specific notion of rurality. The rural which populist leaders refer to is falling behind economically, but often acts as the last bastion of the "true" values of a nation, as well as the "last redoubt" and sole alternative to which to turn in times of crisis, as became clear during the economic crises in countries such as Greece, Portugal and Spain (Döner, Figueiredo and Rivera 2020). It often conserves elements of nationalist imagery and carries promises of identity, juxtaposed against the threats of diffusion and chaos. Whether elements of such rurality really can be found in less-populated settlements outside of big cities is a separate issue. Through its political rediscovery, however, rurality is, willingly or not, placed at the centre of a tempestuous development in which the politics of rural authenticity is coined.

While the political entrepreneurship related to the revival of the rural–urban divide has taken centre stage in contemporary discussions, politics is not the only area through which "authentic rurality" is produced and consumed. In the tourism and heritage industries, arts and popular culture, the idea of rurality is filled with various, often competing, meanings. Rural authenticity is a bestseller, whether you are selling tractors (Brandth 1995), music (Gibson and Davidson 2004) or lifestyle changes (Osbaldiston 2012), but it must be performed and perceived as authentic. The same authenticity, however, plays a key part in the othering of the rural as a periphery (Plüschke-Altof 2016) or as a dangerous place on the margins (Bell 1997; Hayden 2013). In yet another genre of the policies of authenticity, villages and rural areas are being prompted to compete against each other in an effort to produce an image of a "model village", with media attention and prizes awarded to rural "villages of the year" (Pospěch et al. 2015; Kumpulainen 2016). We use the term *policies of authenticity* to summarise these productions, in order to emphasise the deliberate and instrumental use of rurality in these policies, and the active processes whereby various actors fill rurality with different meanings.

The content of this book

In this book, these perspectives are linked to provide a comprehensive view from various national settings on how rural distinctiveness and authenticity are produced, what their functions are and what they mean for actual rural populations. Many of the chapters in this volume are rooted in rural sociology and rural geography, and emphasis is placed on the culturalist approaches within these disciplines, pushing representations, discourses and images of rurality to the forefront of attention. However, these cannot be held separate from actual living conditions in rural areas, and thus we have provided some reflections on the material conditions and developments and the roles they play in producing the policy and politics of authentic rurality. Economic relations within rural areas, as well as those between rural and urban areas, play a key role in our understanding of these developments. Thus, rural underdevelopment can be coded as a sign of backwardness in market-led narratives of progress, but it may also be presented as a carrier of true, unspoilt identities by nationalist movements. Cultural representations stem from the material and economic conditions of life in rural areas, but they also re-frame and transform these conditions and are confronted by them.

This book accounts for the international multiplicity of discursive productions that frame rurality and rural authenticity. Its chapters analyse topics ranging from the Dutch farmers' protests to the Brexit vote, from the integration of immigrants to the politics of the rural idyll, from policy to politics; the volume shows that images of rurality play an important role in contemporary societies and societal conflicts. All of its chapters tackle, from various perspectives, three fundamental questions:

1. How are the notions and images of authentic rurality produced and used in contemporary societies?
2. How did rurality and the rural–urban divide come to be adopted by social movements and populist politicians?
3. What role do the policy- and politics-related images of rurality play in contemporary societies and social conflicts, and how do they relate to the actual lives of rural populations?

These questions speak to fundamental topics in contemporary rural studies, especially rural sociology and rural geography. However, they take the inquiry further, linking the study of rurality with different perspectives and opening it up for different audiences.

This book is organised in two main parts featuring, besides the editors' introduction and conclusion, 11 chapters dealing with the above-mentioned topics and questions. Part I, dealing with the politics of rural authenticity, brings together contributions from the United Kingdom, Estonia, Italy, Norway and Russia. The chapters address a set of populist discourses around rural authenticity, within diverse contexts, ranging from Brexit to neoliberal politics and Covid-19 situations.

The first part of the book opens with the contribution of Jonathan Hearn (Chapter 2) about the polarisation of populism along the rural–urban axis. Based on the UK and US cases and on the premise that while leftist political parties seem to dominate in urban contexts, right-wing parties seem to be dominant in rural areas, Hearn discusses the consequences of this polarisation in the context of nation building. By analysing the shift from the 1970s onwards of the nation-building principles and orientations, due to the expansion of neoliberalism and globalisation processes, he concludes that these dynamics tend to obliterate the rural via the urban and global urbanisation trends. This, Hearn argues, contributes to weakening national integration, as well as to replacing the traditional rural–urban tensions with conflicts around identities and multicultural processes.

In Chapter 3, Michael Woods revisits the politics of the rural by analysing the protests in the British countryside in the early 2000s and their connection to the Brexit process. Woods compares the contexts and motivations of the participants in those protests with the Brexit campaign(ers) to reveal that, despite the few direct connections between both movements, a similar populist rhetoric was used. Furthermore, the distrust in politics that emerged in the rural protests of the 2000s became one of the mottos of Brexit, demonstrating that although the "politics of the rural" may not explain Brexit, it provides important hints regarding the readiness of former rural protesters to join populist movements.

Bianka Plüschke-Altof and Aet Annist, in Chapter 4, examine the connections between rural authenticity and populist movements in the context of neoliberal governance at the regional level in Estonia. They argue that in the context of neoliberalisation of regional politics in this country, rural authenticity is connected to populist movements less via the employment of nationalist discourses and more through the dispossession of those excluded from the rural authenticity regime. Based on a case study, the authors discuss how rural authenticity is used as means of promoting regional development – based on the commodification of rurality – and how it can produce diverse inequalities between rural territories and give way to the expansion of populist discourses and politics.

The tension between the alleged quality and authenticity of Italian agriculture (and food products) and the living conditions of the migrant workers that have recently played a key role in maintaining it is explored by Domenico Perrotta in Chapter 5. That tension seems to have gained new contours in the Covid-19 crisis, with the risk of workforce shortages due to mobility restrictions and the migrant farm workers' vulnerability in the pandemic situation. In response to these risks, a regularisation programme for the inclusion of migrants was implemented by the Italian government. Based on the analysis of the political debate around this programme, mainly focused on government, populist party and union organiser narratives, Perrotta brings to the fore that all three discourses are imbued, to different degrees, with populism. The analysis demonstrates that all these actors use "populism as a method", continuously reminding the audiences of their role and place as connoisseurs of the rural, in order to better legitimise their position in the discussion around Italian agriculture and its workers.

In Chapter 6, Alexander Nikulin and Irina Trotsuk review and discuss the various forms of "populism from above" in Russian regions in respect of rural development, as well as "populism from below", which is strongly anchored in the historical tradition of the "narodnik" movement of community politicisation by the sharing of the same living conditions of local populations. The authors conclude that the latter populism appears to have been revived, while the populism "from above", official and politicised, does not seem to be ideologically convincing in current times. Strongly marked by its historical focus on people, their living conditions and their difficulties, as well as on the peoples' distinctive features from the politicians and authorities, the "narodnik" "populism from below" movement seems different to populist politics elsewhere, and at the same time seems to actively contribute to enhancing and transforming Russian rural communities.

A different perspective on the transformations of rural areas is presented in Chapter 7 by Bjørn Egil Flø, regarding Norway. Flø argues that people in rural Norway feel left behind, and that their communities are dying, as a direct consequence of commodification of "rural authenticity" for tourism development purposes. Traditionally strong Norwegian rural policies, he argues, have been progressively replaced by populist- and neoliberal-related strategies, emphasising entrepreneurship and individual responsibility for rural communities' successful development. Flø then argues that rural people feel betrayed and powerless in the face of such transformations of rural territories.

The second part of the book gathers five chapters that evolve around the policies of rural authenticity, discussing the underlying notions, values and narratives. The section starts with the contribution of Anke Bosma and Esther Peeren (Chapter 8). The authors analyse the Dutch farmers' protests at the end of 2019 (the #Proudofthefarmer movement) and the way in which "rural authenticity" was used as a negotiation currency between protesters and politicians. Bosma and Peeren demonstrate that the long-standing and deeply rooted image of (mainly male) farmers as intrinsically authentic just because *they are* farmers played a significant part in winning sympathetic public opinion, as well as in farmers' alignment with populist-nationalist narratives and politics. Furthermore, the authors argue, farmers' anger and violence during protests was often justified by a shared notion of "rural masculinity".

In Chapter 9, Florian Dünckmann addresses the politics of rural idyll and idyllic politics, referring to the post-political idealisation of rural areas as idyllic places. Dünckmann starts by acknowledging the power of the rural idyll in current societies' imaginary geographies, which have strong argumentative efficacy in political debates. Referring to Hannah Arendt's work, the author argues that populist kitsch narratives convey an image of rural areas as representing an undisputable goodness, as the "right" way of life, that can be found at the basis of deeply anti-democratic and racist ideologies. To illustrate his arguments, mainly how the ideas on the rural idyll can be effective in political debates, he uses two different examples: one related to the memories of a World War II refugee, and the other to the political arguments of proponents of conventional against organic agriculture.

Thoroddur Bjarnason, Brynhildur Thorarinsdottir and Menelaos Gkartzios discuss, in Chapter 10, the representations of rurality in Icelandic cinema as the opposite of the idyll. Starting from the presentation of Icelandic cinema as a vibrant industry, the authors highlight the fact that the majority of award-winning films are set in rural and remote communities. These awards mainly recognise cinematographic representations of the "real rural", a dystopic Icelandic rural territory – gloomy, violent, wild and harsh. The authors argue that this commodification of rural dystopia legitimises urban-centred social and regional policies, while supporting the success of the film industry, and mainly of urban film-makers.

In Chapter 11, by Guro Korsnes Kristensen and Berit Gullikstad, immigrants' integration in Norwegian villages is examined. The authors explore the ways in which experiences with immigration and immigrant integration are narrated, highlighting the place of "local authenticity" in those experiences and discourses. The analysis indicates the reframing of "rural authenticity" into cosmopolitism and heterogeneity. In fact, rural areas seem to do better regarding immigrant integration than their urban counterparts, by providing a stable, strong and authentic environment. The authors conclude that narratives on immigration and integration can contribute to the production of rural authenticity, and therefore also to the rural–urban debate.

Grzegorz Foryś, in Chapter 12, analyses the dynamics of change in farmers' protests in Poland during the last three decades, demonstrating its increasing institutionalisation and organisation, far removed from the protests of the post-socialist era. Nowadays, protests are less frequently used by farmers as a way of claiming their own rights. Rather, Foryś notes, protests are used as actions to support farmers' activity in the institutionalised sphere. This change seems to correspond to the abandonment of peasants' traditions and symbols in the protests, and to their replacement by nationalist and religious signs. All these changes have transformed the relationships between farmers and the state, paving the way for a model of institutionalised cooperation.

The book ends with a discussion, by the editors, on the main dimensions of the politics and policies of authenticity nowadays. These are, indeed, manifold and multiple, as their combinations and manifestations and seem to be strongly anchored in territorial, historical and political contexts. The multiplicity and complexity of the topics covered in this book clearly show the demand for further research, regarding both the examination of other dimensions of the politics and policies of rural authenticity and, perhaps more importantly, their other contexts.

The editors believe that the chapters gathered in this volume provide a timely analysis of the politics and policies of rural authenticity, and of the ways in which these are interwoven with the current wave of national populism that is sweeping across Europe and the USA. With this, the editors would like to thank all the authors for their contributions to this volume, as well as the reviewers of the volume. We also extend our thanks to Faye Leerink, Nonita Saha and the Routledge editorial team.

Note

1   Michel Houellebecq's book *Serotonin* (Houellebecq 2019) provides an interesting literary account of the current rural and peripheral unrest in France.

References

Bell, D., (1997) Anti-idyll: Rural horror. In Cloke, P., Little, J. (eds.), *Contested countryside cultures: Otherness, marginalisation and rurality*. London, Routledge, pp. 94–108.

Bell, D. (2006) Variations on the rural idyll. In Cloke, P., Marsden, T., Mooney, P. H. (eds.), *Handbook of rural studies*. London, Sage, pp. 149–160.

Bell, M. (1992) The fruit of difference: The rural-urban continuum as a system of identity, *Sociologia Ruralis*, 57(1), pp. 65–82.

Brandth, B. (1995) Rural masculinity in transition: Gender images in tractor advertisements, *Journal of Rural Studies*, 11(2), pp. 123–133.

Brubaker, R. (2020), Populism and nationalism, *Nations and Nationalism*, 26(1), pp. 44–66.

Buttel, F. and Newby, H. (1980) *Rural sociology of the advanced societies*. Allanheld, Osmun.

Cloke, P. (2006) Conceptualizing rurality. In Cloke, P., Marsden, T., Mooney, P. H. (Eds.), *Handbook of rural studies*. London, Sage, pp. 18–27.

Cramer, K. (2016) *The politics of resentment: Rural consciousness in Wisconsin and the rise of Scott Walker*. Chicago and London, The University of Chicago Press.

Döner, F., Figueiredo, E. and Rivera, M. J. (2020) (eds). *Crisis and post-crisis in rural territories: Social change, challenges and opportunities in Southern and Mediterranean Europe*. Dordrecht, Springer.

Eatwell, R. and Goodwin, M. (2018) *National populism: The revolt against liberal democracy*. London, Pelican.

Figueiredo, E. (2013) McRural, no rural or what rural? – Some reflections on rural reconfiguration processes based on the promotion of Schist villages network, Portugal. In Silva, L., Figueiredo, E. (eds.) *Shaping rural areas in Europe: Perceptions and outcomes on the present and the future*. Dordrecht, Springer, pp. 129–146.

Friedland, W. H. (1982) The end of rural society and the future of rural sociology. *Rural Sociology*, 47(4), pp. 589–608.

Fukuyama, F. (1992) *The end of history and the last man*. New York, The Free Press.

Fukuyama, F. (2018) *Identity: The demand for dignity and the politics of resentment*. New York, Farr, Straus and Giroux.

Gellner, E. (2006) *Nations and nationalism*. Oxford, Blackwell Publishing.

Gibson, C. and Davidson, D. (2004) Tamworth, Australia's 'country music capital': Place marketing, rurality, and resident reactions. *Journal of Rural Studies*, 20(4), pp. 387–404.

Goodhart, D. (2017) *The road to somewhere: The populist revolt and the future of politics*. London, C. Hurst and Co ltd.

Guardian. (2019) Fight for Europe or the Wreckers will destroy it, 25 Jan., downloaded 01.04.2019 www.theguardian.com/commentisfree/2019/jan/25/fight-europe-wreckers-patriots-nationalist

Guilluy, C. (2019) *Twilight of the elites: Prosperity, the periphery and the future of France*. London, Yale University Press.

Halfacree, K. (1993) Locality and social representation: Space, discourse and alternative definitions of the rural. *Journal of Rural Studies*, 9, pp. 1–15.

Halfacree, K. (2006) Rural space: Constructing a three-fold architecture. In Cloke, P., Marsden, T., Mooney, P. H. (eds.) *Handbook of rural studies*. London, Sage, London, pp. 133–148.

Halfacree, K. (2007) Trial by space for a "radical rural": Introducing alternative localities, representations and lives. *Journal of Rural Studies*, 23, pp. 44–63.

Hall, J. A. (2019) Our current sense of anxiety after Gellner. *Nations and Nationalism*, 25(1), pp. 45–57.

Harris, E. (2016) Why has nationalism not run its course? *Nations and Nationalism*, 22(2), pp. 243–247.

Hayden, K. (2013) Inbred horror: Degeneracy, revulsion, and fear of the rural community. In Fulkerson, G., Thomas, A. (eds.), *Studies in urbanormativity: Rural community in urban society*. New York, Lexington Books, pp. 181–206.

Houellebecq, M. (2019) *Serotonin*. Farrar, New York, Straus and Giroux.

Hosking, G. (2016) Why has nationalism revived in Europe? *Nations and Nationalism*, 22(2), pp. 210–220.

Kumpulainen, K. (2016) The village of the year competition constructing an ideal model of a rural community in Finland. *Sociální Studia/Social Studies*, 13(2), pp. 55–71.

Mounk, Y. (2018) *The people vs democracy: Why our freedom is in danger and how to save it*. London, Harvard University Press.

Osbaldiston, N. (2012) *Seeking authenticity in place, culture, and the self: The great urban escape*. London, Palgrave.

Pahl, R. (1966) The rural-urban continuum. *Sociologia Ruralis*, 6(3), pp. 299–329.

Piketty, T. (2014) *Capital in the twenty first century*. Cambridge, Harvard University Press.

Piketty, T (2020) *Capital and ideology*. Cambridge, Harvard University Press.

Plüschke-Altof, B.(2016) Rural as periphery per se? Unravelling the discursive node. *Sociální Studia/Social Studies*, 13(2), pp. 11–28.

Pospěch, P., Spěšná, D. and Staveník, A. (2015) Images of a good village: A visual analysis of the rural idyll in the "Village of the Year" competition in the Czech Republic. *European Countryside*, 7(2), pp. 68–86.

Rodríguez-Pose, A. (2018) The revenge of the places that don't matter (and what to do about it). *Cambridge Journal of Regions, Economy and Society*, 11(1), pp. 189–209.

Rodríguez-Pose, A. (2020) The rise of populism and the revenge of the places that don't matter. *LSE Public Policy Review*, 1(1): 4, pp. 1–9.

Rokkan, S (1987) *Stat, nasjon, klasse*. Oslo, Universitetsforlaget.

Simmel, G. (1950) The Metropolis and mental life. In Wolff, K. (ed.) *The sociology of Georg Simmel*. Glencoe, Free Press, pp. 409–424.

Soares da Silva, D., Figueiredo, E., Eusébio, C. and Carneiro, M. J. (2016) The countryside is worth a thousand words – Portuguese representations on rural areas. *Journal of Rural Studies*, 44, pp. 77–88.

Sorokin, P. A. and Zimmerman, C. C. (1929). *Principles of rural-urban sociology*. New York, Henry Holt and Company.

Wirth, L. (1938) Urbanism as a way of life. *American Journal of Sociology*, 44, pp. 1–24.

Part I
Politics of rural authenticity

2 City and countryside in the imagining of nations

Jonathan Hearn

Introduction

When we look at maps illustrating the distributions of support for Donald Trump versus Hillary Clinton in the 2016 US presidential election,[1] and of support for "Leave" versus "Remain" in the UK Brexit referendum in 2016, we see a striking similarity. Support for Clinton and for Remain both appear as archipelagos of major cities and surrounding urban concentrations, poking up out of a sea of support for the opposite causes. City and countryside seem to define patterns of polarisation in both societies. Of course, these images indicate majorities in constituencies; the polarisation of opinion and support is more gradual than the image of "above and below the sea line" suggests. Nonetheless, the underlying social tension, and its spatial expression, are real. In this chapter I argue that this can be seen as a symptom of the evolving problem of nation building, which has itself been altered by changing relations between city and countryside. Neoliberalism and globalisation have changed the terms of the relationship, in some ways eliding it completely. Yet the central task of nation building, confronted by spatialised social tensions, persists. The currency of new notions of multiculturalism, and revived notions of civil society, reflect an increasingly urban-centric process of nation building, which nonetheless must re-forge relations with its hinterland.

I explore the rural–urban divide with regard to the idea of the nation and the recent rise of populist forms of politics in the early 21st century. For the latter, I will limit my speculations to the two iconic cases I am more familiar with: support for Donald Trump's presidency in the US, and for the project of Brexit in the UK. However, I am aware that the phenomenon is much more widespread and varied, and difficult to disarticulate from top-down projects of autocratic rule (e.g. the governments of Erdogan in Turkey, Bolsonaro in Brazil and Orbán in Hungary). Nonetheless, I will try to place these two cases of populism, and their relationship to rural–urban tensions, in a longer historical context of "nation building" (Deutsch, 1953; Bendix, 1996; Wimmer, 2018) and how that has been altered by the political and economic changes stemming from the 1970s. I present this as two subsequent historical narratives: the first a period of classic nation building associated with the rise of the modern, centralised bureaucratic state, the second a

DOI: 10.4324/9781003091714-3

period of what I will call "nation deconstructing" associated with the recent period of economic globalisation and the hegemony of neoliberal policies (Harvey, 2005; Mann, 2013: 129–178). Before setting up this contrast, let me do some basic framing with regard to theories of nationalism and my two cases of populism (Trump and Brexit).

Gellner (1983), that seminal figure in the post-1970s study of nationalism, was famous for treating it as hinging on a great historical transition from agrarian empires to industrial states: from territories of relatively isolated peasant communities ruled by narrow urban elites, to societies where all are much more integrated into a dynamic market economy, in which peasants have been transformed into mobile workers and elite culture generalised to mass populations. In his celebrated historical study, Eugen Weber (1976) treated France as a type-case of the general process, in which developments in communications, education, linguistic standardisation and militarily driven social mobility transformed "peasants into Frenchmen". These are just two expressions of a very general paradigm that has informed understandings of the rise of nationalism as involving the impact of the urban on the rural. However, Gellner's version in particular is very broad brushed, and needs to be qualified in at least two ways. First, as always happens with such ideal-typical abstractions, the reality is not so clear cut. For instance, the United States was still a largely agricultural economy well into the early 20th century, with almost 40% of employment being in agriculture as late as 1910. The UK was peculiar in abandoning food self-sufficiency quite early as part of its pre-cocious industrialisation (and France was somewhere in between) (Roser, 2013). We should think of the process that Gellner was highlighting not as a sharp shift from rural to urban, or from agrarian to industrial, but as a changing relationship between these, in which the city was becoming the more dynamic partner when it came to generating employment. Second, Gellner's model tended to assume that once the agrarian-industrial transition had happened, the modern nation had been constructed and the process was complete. The nation becomes part of the taken-for-granted furniture of modern life. I would argue instead that nation building is perennially unfinished business, and that the legitimacy of the system of rule needs to be constantly renewed, as developments redistribute power and opportunity among society's members. Indeed, the liberal democratic form is precisely premised on the idea of a permanent debate about who "the people" are, what their shared values are and how they should govern each other (see Hearn, 2006: 165–169, *passim*). This view fundamentally informs what follows.

Now for Trump and Brexit. Here the main qualification is that we need to be careful about what we mean by "rural" and "urban". The polarisations of Trump versus Clinton, and Leave versus Remain, have been summarised through this shorthand, but it can be misleading in two ways, which I will expand upon in what follows, but outline briefly here. First, the "rural" and the "urban" are not what they used to be. Not only are sources of food much more transnational than they used to be (the US imports about 15% of its food, the UK just under half), but the replacement of agricultural labour by technology has drastically changed the structure of employment outside of big cities. Rural economies tend to combine low-employment agriculture with some support industries, government-sponsored projects such as forestry and, crucially in many places, seasonal tourism and its various "attractions" and support services. In turn, major cities are no longer the hubs of major industry they once were. Instead, they are centres of regional government, finance and financial services, and myriad other private and public services and retail activities. Much industrial activity, especially in the capitalist core, has been relocated either abroad or into smaller, mid-sized towns, as part of projects of economic renewal. So, in the popular discourse about a new rural–urban divide contributing to populism, those many ailing, mid-sized, semi-industrial towns are now often included in what we mean by "rural". Second, and perhaps more simply, it has rarely been an equal partnership. Power concentrates in cities, which amass and coordinate resources, and have since their inceptions dominated their hinterlands (Tilly, 1992: ch. 1). The basic point here is that the language of rural and urban, which can be applied over millennia of human history, can mask substantive changes of composition. Attending to those changes will help us get some analytic purchase on the "rise of populism".

Classic nation building

To avoid getting too diffuse, I will advance a general argument on the basis of my two cases (the UK and the US), allowing that further comparisons would probably complicate and modify my argument. Here and in the next section, my account will move from the political economy of the rural–urban relationship to the cultural representations of that relationship.

Various aspects have been emphasised in the story of the British nation formation: Protestant identity, recurrent wars with (Catholic) France, global imperial conquest and liberal economics (Colley, 2009; Kumar, 2015). But here let me focus on the question of integrating Britain, city and countryside, as a nation. In Book III of his *Wealth of Nations* (1981: 376–427), Adam Smith (1981) made the rural–urban relationship central to his historical analysis of how Europe, especially England/Britain, managed to develop production, pulling itself out of the historical trap of cycles of agricultural crisis, by generalising prosperity and consumption to wider populations. Smith was keen to argue against a theory, associated with the French *Economistes* or "Physiocrats", that maintained a fairly strict hierarchy of economic development, with agriculture seen as the foundation on which first manufactures and then trade emerge. He argued that Europe's "unnatural" (1981: 380) path of economic growth worked in the other direction. It was driven by semi-autonomous towns and cities, preserved from the Roman period, that kept manufacturing and trade alive through the Middle Ages, eventually expanding as commerce grew, in turn generating pools of capital that were invested back into agricultural improvement, causing population growth and further growth in manufactures and trade. It is this virtuous circle of economic growth between city and countryside that first underwrote the growth of the evolving core nation-states in the modern period. For generations the rural hinterlands of modern nation-states thrived on supplying food to cities, shedding excess populations into cities and imperial outposts. And, in turn, cities supplied growing markets and sources of further capital investment.

Through the 19th and most of the 20th centuries this reciprocal relationship was accompanied by the institutional integration of the rural and the urban. National systems of schooling spread across the landscape, often more effectively for the denizens of small agricultural towns than for the urban labouring poor. Rural dialects were assimilated to and measured against dominant urban forms. Standardisation of curricula and narratives around such things as national history helped shape a more unified worldview. Older systems in which rural magnates and elites operated local systems of law in their own interests were increasingly subsumed under national regimes of law. And successive waves of advances in communications – roads and canals, postal systems, telegraphs and telephones, news services – pulled entire regions together, not effacing the rural–urban divide, but giving each its place in a major axis of the division of labour (Deutsch, 1953).

As already suggested, across this period class structure was steadily changing. As a proportion of sectoral employment, agriculture in Britain had reduced to a smaller overall proportion than manufacturing by 1800, whereas in the USA the growth of manufacturing employment did not begin to equal and overtake agriculture until the early 20th century. However, over the long term, up to the present, both were steadily displaced as a proportion by the "service sector" (a very broad category), which is now about 80% in both countries, the vast bulk of the remainder of employment being in manufactures (Ortiz-Ospina & Lippolis, 2017). But for our first period of "nation building", up to about 1970, boosted by post–World War Two economic expansion, demographic flow from the shrinking sectors into the growing service sector supported a sense of prosperity and upward mobility. A new kind of larger and more heterogeneous middle class took shape in both countries, associated with this growth of the diverse service sector. This was correspondingly associated with the steady movement of populations from the countryside into the cities (ibid.).

Many aspects of high and popular culture from the late-19th to the mid-20th century reflected attempts to grapple with the rural–urban relationship. Familiar, dualistic conceptual mainstays of early sociology are a case in point. Henry Maine's (1986) legal studies of a shift from "status" to "contract", Ferdinand Tönnies' (2001) contrasts of Gemeinschaft and Gesellschaft, and Emile Durkheim's (1964) shift from "mechanical" to "organic" solidarity all attempted to grasp aspects of the rural–urban relationship, and the social disruptions of movement from one to the other. These ideas were generally framed as describing a longer historical shift from "traditional" to "modern" society, but their salience for these theorists and their audiences surely had to do with a sense that they were not just historical but currently relevant, as people could see the tensions and adjustments between these principles all about them.

More popular forms of culture also expressed this relationship. For instance, in Scotland, where this shift was experienced perhaps a bit later than in the south of England, there developed a specific form of popular literature known as the "Kailyard School" (1880–1914). A particular variant on the *Bildungsroman*, these stories characteristically featured the trials and tribulations of a talented and promising young man from a small "toun" as he made his way in the modern city, only to return home, wiser for the experience, and newly appreciating the values and relationships of home (Nash, 2007). This attempt to accept the rural–urban relationship and revalorise while sentimentalising the rural can be found in many other national variants on this pattern. The present author remembers from his childhood in America a diet of reruns of 1960s television shows that revolved around this same dilemma of how the city perceived the countryside. Popular shows such as "The Beverly Hillbillies", "Green Acres" and "The Andy Griffith Show" found humour in playing on an ambiguous combination of condescension and sentimental admiration for the moral and practical good sense of rural folk, uncorrupted by urban sophistication. This could only make sense as a brand of popular humour for mass consumption if it was generally recognisable. And indeed, the middle years of the 20th century in the USA especially were ones in which many of the urban viewers could remember the way of life being parodied and sentimentalised, either through their own life trajectories or via family connections and memories. These narratives provided a way of reaching across a social divide and processing a sense of a vanishing way of small-town, agriculturally based life. It was a means of constructing a shared national identity, even as its composition was fundamentally changing.

Nation deconstructing

In the last 50 years, we have moved through a new period in which manufacturing sectors have gone into decline, and in some places crisis, while employment in entire societies has been almost swallowed up by the capacious category of the "service sector", which contains and occludes profound degrees of social stratification, from high-end financial specialists, high-tech experts, and professionals (doctors, lawyers, scientists, professors) to low-end precarious jobs in hospitality, cleaning, delivery and transportation. This restructuring corresponds to economic globalisation, in which many manufacturing jobs have moved away from the USA and the UK, and as legal and illegal labour migration has helped undercut the bargaining power of these transformed working classes in the old core. The Smithian division of labour between town and countryside has been undercut, with agriculture providing very little employment and agricultural products being bought and sold on much more international markets (Herrendorf, Rogerson, & Valentinyi, 2014).

The schooling of the previous period has rightly been criticised for its myth-making representations of the forming nation and its many exclusions from, or biased accounts of, history. In the USA the idea of the "melting pot" reigned (Herberg, 1960; Glazer & Moynihan, 1970), in which diverse religious and ethnic groups were assimilated to a core national identity modelled on a white, British and northern European standard. The continuing legacy of slavery and racism were obscured by a narrative of post–Civil War reconciliation between the North and South and romanticising of the South's "Lost Cause". Fear of atheistic communism in the USSR and its clients further unified society in the post–World War Two years (Katznelson, 2013). More recent decades, in the wake of the 1960s civil rights movements in both the USA and the UK, have seen a salutary critique of much of this earlier nation-building history, focusing on the many patterns of discrimination and exclusion that underlay this social order. While regionally varied and often contested, especially in more urban and progressive school districts, an approach to national history that was more a matter of collective soul-searching in regard to historical injustices, rather than national mythologising, became much more normal (see, e.g., Hannah-Jones, 2019; Magness, 2020). In recent years, young people have been increasingly socialised into a very different vision of the social landscape – one characterised by problems of disunity and injustice within the nation, in which identity politics is a major frame of interpretation and controversial ideas of "political correctness" and "wokeness" are at stake. This new narrative carries on into much liberal college and university education, which, as we know, is one of the key experiences distinguishing Trumpians from Clintonians, and Leavers from Remainers (Silver, 2016; Swales, 2016).

Patterns of communication have also been altered. Benedict Anderson (2006) argued that the spread of literacy and the printed word through new markets of reading publics was crucial to the original imagining of modern nations as communities. As suggested above, by the middle of the 20th century broadcast media were generating regional and national diets of programming on radio, and then television, that provided a common focus for the public imagination. Agencies such as the BBC saw themselves as having a mission of national education and uplift. However, the rise of personal everyday computing power along with satellite-based communications has fundamentally changed communicational dynamics. Rather than the mass of society "receiving" shared, nationally framed images and messages, entertainment viewing, whether on Netflix or YouTube or some other content deliverer, is on demand, according to personal taste. As has often been observed, email and countless social media platforms individualise paths of communication, as these technologies also serve the purpose of generating bespoke market information to be sold to third parties (Turkle, 2017). All the while, the print news media, once one of the cornerstones of national communication, has gone into steady decline (Rusbridger, 2018). Moreover, we shouldn't forget the other meaning of "communication" as the movement of people and things. The rise of the automobile facilitated the circulation of people, but largely within the nation-state horizons, visiting sites of recreation and national beauty, and relatives "back home" in small towns. With the cheapening and expansion of air travel, however, the middle classes are now more likely to travel abroad for tourism and recreation, dropping in briefly on other cultures, perhaps as part of a group of like-minded friends, rather than circulating within the national sphere, traversing the urban–rural divide.

Social mobility and the growth of the middle class have stalled in this period. This is partly because in previous decades social mobility was largely measured in terms of movement from "blue collar" manufacturing work to "white collar" professional/service sector work (Goldthorpe, 2016: 95–96). However, as the vast majority of employment is now in the highly variegated service sector, this pattern is largely exhausted. As already suggested, major class divisions now exist within the service sector. Add to this the erosion of working conditions and job security associated with the decline of trade union power, the stagnation of middle-class incomes and the meteoric rise in incomes at the very top of the wealth hierarchy, and it becomes clear that participation in an expanding national economy is no longer the unifying project it once was. While inequality does strain the social fabric, this is not just a matter of wealth distribution, but also of a sense of inclusion in a shared world of valued and stable employment. The effects of this process are compounded by national housing markets increasingly out of the reach of young people ready to set up households, especially since the property-market crash of 2008. To be sure, this basic shift in class structure and dynamics cuts across the rural–urban divide. Ethnically diverse working classes in the growing cities experience precarity as much as underemployed rural workers in stagnating small towns. Again, the difference is not so much one of class in the broadest sense as one of experience and perception, and of orienting values. But these latter are nonetheless connected to how one sits in the overall division of labour, and one's sense of one's interests and prospects in regard to the wider economy (Cramer, 2016; Goodhart, 2017; Lind, 2020).

With regard to popular representations of the rural–urban relationship, what is striking is not so much an axis of hostility as the relative disappearance of the symbolic relationship compared to the previous nation-building period. Obviously, there are all kinds of real and symbolic tensions expressed in the social polarisations captured by Trump's presidency and Brexit, but, as the polls tell us, these are about several things: age, levels of education, race and ethnicity, religion and so on, not just the rural–urban axis. The supposed rural–urban tensions have become visible through patterns of voting behaviour, catching many observers off guard. But what is noticeable in popular culture is the elision of a tension that was routinely represented, however sympathetically, in the earlier period. The rural appears as a place of natural beauty, archaeological interest, passing attention, but not so much as a way of life, vanishing or not. One exception in the UK is the long-running radio drama *The Archers* on Radio 4, a station whose listenership leans heavily towards the urban, well-educated, Remain end of the spectrum. The Archers, first broadcast in 1951, follows the ups and downs of various farming families in the imaginary village of Ambridge, and was originally formulated as a show with a specific remit to disseminate important information about rural farming life to a wider public. In other words, it was a product of that earlier conjunction of nation building and broadcast media. However, today's listeners are mostly very distant from a farm-based or rural lifestyle.

Returning to the realm of television drama and comedy, while representations of urban working-class people are not uncommon, small-town life as a frame for humour or drama, while not absolutely absent, is increasingly rare, especially if the setting is present day. The rural is usually seen as a place that urban protagonists (e.g. police detectives) must go to sometimes in the interests of plot. On the one hand, there is little representation of the rural–urban relationship as such; on the other, for all the reasons of current means of communication discussed above, the difference itself is elided. The rural today is no longer an "other" place, the symbolic opposite of the city; it is more the far outer margins of the city. Everyone has a cell phone, a computer, a car, watches Netflix. In a sense, there is only one way of life left (cf. Buttel, 2001; Fulkerson & Thomas, 2013). But some people seem to live their lives close to the centre of it, partaking of the city's dynamism and prestige, while others are located on its periphery, in a slightly dreary, downtrodden side-life, where people can still be stalwart and admirable, but are seldom "where it's at". As discussed above, the options for employment and housing in rural areas are different, on the whole, and people clearly identify on average in more socially conservative ways, so there are variances. But this is more a difference in the social and economic prospects facing people in major metropolitan centres versus declining towns and rural areas than between two "cultures": one urban, the other rural.

Discussion

Having framed this as a contrast between two periods – one "classic nation building", the other "nation deconstructing" – I might be understood to be supporting the kinds of arguments that were prevalent in the early days of the "globalisation" literature, in which the "end of the nation state" was breezily forecast (e.g. Held, 1996; cf. Hall, 2000). But that is far from my position. The global nation-state system, and thus the perennial problems of nation building, are built deeply into the world's political economy, something that is very evident in the nationally framed responses to the coronavirus crisis that have shaped the global pandemic at the time of writing this chapter. Predictions of the nation-state's death are grossly exaggerated. However, as the economies and political power balances within states evolve, the basic terms of nation-building shift. As labouring populations outside of major cities (agriculture, but also mining and other industries) have been whittled down and largely replaced with a rural wing of the service sector, and urban industrial workers have been rechannelled into myriad jobs around the knowledge economy, commerce, and consumption, an older sense of a reciprocal (though sometimes fraught) relationship between city and countryside, as parts of a unified national division of labour, has decayed. It is no longer an object, or problematic, of nation building. Meanwhile, that problematic has migrated into a different frame, more concerned with how to imagine an ethnically, racially and religiously diverse nation without invoking a strong principle of assimilation. This frame also attends to the membership rights of both sexes, and various sexualities and gender identities. Thus, the narrative of nation building has not ended, but its focus has shifted to the terrain of multiculturalism, to a different problem of how to build a unified nation (Modood & Werbner, 1997; Barry, 2002; Parekh, 2002).

In this context of what I am describing as a new multicultural nation-building project, some efforts to make sense of the Trump and Brexit phenomena have seen it primarily as a reactionary expression of resurgent racism, a backlash against feminism and the promotion of women's rights and, in the UK, as a kind of British imperial nostalgia that seeks to turn back the clock to a more ethnically and racially homogenous society (Bhambra, 2017; Patel & Connelly, 2019). This has led to a rather unproductive opposition of explanations based on "racism", as an expansive ideology, versus "class", as a complex of economically grounded interests. Clearly both racism and class interests are involved, but neither alone provides a magic key to unlock the causes of these events. As I've tried to suggest throughout this chapter, the rural–urban divide cannot be reduced to a simple class divide (and perhaps less so than in the past). Yet there are interests, and senses of identity, that attach to places and the livelihoods they afford, and there is a palpable pattern of alienation in many rural areas that feel relatively stagnant. It is also the case that ethnic and racial diversity is concentrated in cities, and to some degree this becomes a marker in rural perceptions of the urban. However, that doesn't make race the central issue. I suggest that as cities have continued to grow, the endless project of nation building has shifted its focus from the rural–urban relationship to more characteristically urban problems of national integration. Rural and small-town folk have been not so much "left behind" as marginalised in the unavoidable and ever-problematic discussion of "who we are" as a nation.

This chapter aims to consider the onward march of "urbanisation", not only in the narrow sense of cities growing and people moving into them, but also in the wider sense, concomitant with globalisation, of the entire world becoming one big city. Despite profound divisions of political and economic interests shaped by territorial states, the traffic of commodities and ideas between major global cities increasingly foreshadows one huge macro-city that girdles the globe (see Brenner, 2019; Ritchie & Roser, 2019). It is a global space in which elites and the more affluent strata on the one hand, and work-seeking and politically displaced populations on the other, have circulated to unprecedented degrees. Driven by globalisation and neoliberalism, this global network now overlays the system of nation-states. But that doesn't mean it can erase that system, because those states are still the main framework within which people make claims on each other about basic social provisioning, and moral and legal obligations and rights. It is still the main framework within which they mobilise around interests and contest distributions of power. This is perhaps why we see the current wave of national populisms around the globe, each with its own particular variant of tensions between "rural" and "urban" interests.

In recent decades, the concept of "civil society" has been revived and extended (Hall, 1995; Edwards, 2004): revived as a framework for thinking about social mobilisation around issues and interests, and "progressive" social change, and extended to the global domain to account for the role of various international bodies that operate at the interstices of states, supposedly helping to forge a genuinely global society of shared interests. Achieving this latter goal still seems a long way off. Yet, in the present context it is perhaps instructive to meditate on the long-standing affiliations of the concept with the "civilising" effects of the city and commerce, bringing to heel more rural and martial ways of life, and establishing the rule of law across the land. In the 18th century, before modern democracy was fully conceived, and before dynastic and imperial states had evolved into more recognisable nation-states, this was the original nation-building project: bringing the countryside under the benevolent influence of the city (a process regarded with some regret by Adam Ferguson (1966) in 1767). From the late 19th to the mid-20th centuries, in the heyday of classic nation building, the term fell out of fashion in elite discourse, replaced by languages of class conflict (Marx) and the functional integration of society (Durkheim), understood at least implicitly to be problems of the nation-state, or at least practically focused at that level. The return of "civil society" – now often "global civil society" (Kaldor, 2003) – is perhaps a symptom of the global urbanisation referred to above. And although the domination of urban life may be inexorable and bring with it some values we want to promote, it's perhaps not surprising that the somewhat condescending civilising mission of the city in regard to the countryside continues to antagonise people who feel themselves marginal to that centre. But, as I've tried to suggest throughout this chapter, given the transformations of that relationship, and the deep and global interconnections of contemporary capitalism, perhaps we are talking about people who are not so much "rural" as encamped just outside the city walls, and with more in common than we realise with those struggling to survive within those walls. Those living in the castle, however, are much more distant from all the rest.

Note

1 The 2020 US Presidential election was held after this chapter was written. However, its results reaffirm the spatial polarisation of party support discussed here.

References

Anderson, B. 2006. *Imagined communities: Reflections on the origin and spread of nationalism.* Revised edn. London: Verso.

Barry, B. 2002. *Culture and equality: An egalitarian critique of multiculturalism.* Cambridge, MA: Harvard University Press.

Bendix, R. 1996. *Nation building and citizenship: Studies of our changing social order.* 3rd edn. Piscataway: Transaction Publishers.

Bhambra, G. K. 2017. Brexit, Trump, and "methodological whiteness": On the misrecognition of race and class. *British Journal of Sociology*, 68(1), pp. 214–232.

Brenner, N. 2019. *New urban spaces: Urban theory and the scale question.* Oxford: Oxford University Press.

Buttel, F. H. 2001. Some reflections on late twentieth century agrarian political economy. *Cadernos de Ciência & Tecnologia*, 18(2), pp. 11–36.

Colley, L. 2009. *Britons: Forging the nation 1707–1837.* 3rd edn. New Haven: Yale University Press.

Cramer, K. J. 2016. *The politics of resentment: Rural consciousness in Wisconsin and the rise of Scott Walker.* Chicago: University of Chicago Press.

Deutsch, K. W. 1953. *Nationalism and social communication.* 2nd edn. Cambridge, MA: MIT Press.

Durkheim, E. 1964. *The division of labor in society.* New York: Free Press.

Edwards, M. 2004. *Civil society.* Cambridge: Polity Press.

Ferguson, A. 1966/1767. *An essay on the history of civil society.* Forbes, D. (ed.). Edinburgh: Edinburgh University Press.

Fulkerson, G. M. and Thomas, A. R. (eds) 2013. *Studies in urbanormativity: Rural communities in urban society.* Lanham: Lexington Books.

Gellner, E. 1983. *Nations and nationalism.* Cornell: Cornell University Press.

Glazer, N. and Moynihan, D. P. 1970. *Beyond the melting pot: The Negroes, Puerto Ricans, Jews, Italians and Irish of New York city.* 2nd edn. Cambridge: MIT Press.

Goldthorpe, J. H. 2016. Social class mobility in modern Britain: Changing structure, constant process. *Journal of the British Academy*, 4, pp. 89–111.

Goodhart, D. 2017. *The road to somewhere: The populist revolt and the future of politics.* London: Hurst and Company.

Hall, J. A. (ed) 1995. *Civil society: Theory, history, comparison.* 2nd edn. Cambridge: Polity.

Hall, J. A. 2000. Globalization and nationalism. *Thesis Eleven*, 63, pp. 63–79.

Hannah-Jones, N. (ed) 2019. The 1619 project. *New York Times Magazine*, August 14, 2019. Retrieved from: www.nytimes.com/interactive/2019/08/14/magazine/1619-america-slavery.html.

Harvey, D. 2005. *A brief history of neoliberalism.* Oxford: Oxford University Press.

Hearn, J. 2006. *Rethinking nationalism: A critical introduction.* Basingstoke: Palgrave Macmillan.

Held, D. 1996. The decline of the nation state. In Eley, G. and Suny, R. G. (eds), *Becoming national: A reader.* Oxford: Oxford University Press, pp. 407–417.

Herberg, W. 1960. *Protestant, Catholic, Jew: An essay in American religious sociology.* Chicago: University of Chicago Press.

Herrendorf, B., Rogerson, R. and Valentinyi, A. 2014. Growth and structural transformation. In Aghion, P. and Durlauf, S. N. (eds), *Handbook of economic growth, Vol. 2.* London: Elsevier, pp. 855–941.

Kaldor, M. 2003. *Global civil society: An answer to war.* Cambridge: Polity.

Katznelson, I. 2013. *Fear itself: The new deal and the origins of our time.* New York and London: Liveright.

Kumar, K. 2015. *The idea of Englishness: English culture, national identity and social thought.* London: Routledge.

Lind, M. 2020. *The new class war: Saving democracy from the metropolitan elite.* London: Atlantic Books.

Magness, P. W. 2020. *The 1619 project: A critique*. Creative commons. The American Institute for Economic Research.

Maine, H. 1986/1861. *Ancient law*. USA: Dorset Press.

Mann, M. 2013. *The sources of social power, vol. 4: Globalizations 1945–2011*. Cambridge: Cambridge University Press.

Modood, T. and Werbner, P. 1997. *The politics of multiculturalism in the new Europe: Racism, identity, and community*. Basingstoke: Palgrave Macmillan.

Nash, A. 2007. *Kailyard and Scottish literature*. Amsterdam and New York: Rodopi.

Ortiz-Ospina, E. and Lippolis, N. 2017. Structural transformation: How did today's rich countries become "deindustrialized". *Published online at* OurWorldInData. org. Retrieved from: https://ourworldindata.org/structural-transformation-and-deindustrialization-evidence-from-todays-rich-countries.

Parekh, B. C. 2002. *Rethinking multiculturalism: Cultural diversity and political theory*. Cambridge, MA: Harvard University Press.

Patel, T. G. and Connelly, L. J. 2019. "Postrace" racism in the narratives of "Brexit" voters. *The Sociological Review*, 67(5), pp. 968–984.

Ritchie, H. and Roser, M. 2019. *Urbanization. Published online at* OurWorldInData. org. Retrieved from: https://ourworldindata.org/urbanization.

Roser, M. 2013. Employment in agriculture. Published online at *OurWorldInData.org*. Retrieved from: https://ourworldindata.org/employment-in-agriculture

Rusbridger, A. 2018. *Breaking news: The remaking of journalism and why it matters now*. Edinburgh: Canongate.

Silver, N. 2016. *Education, not income, predicted who would vote for Trump*. Published online at *FiveThirtyEight.com*. Retrieved from: https://fivethirtyeight.com/features/education-not-income-predicted-who-would-vote-for-trump/

Smith, A. 1981/1776. *An inquiry into the nature and causes of the wealth of nation. 2 vols*. Indianapolis: Liberty Fund.

Swales, K. 2016. *Understanding the leave vote*. London: NatCen Social Research.

Tilly, C. 1992. *Coercion, capital, and European states, AD 990–1992*. Oxford: Blackwell.

Tönnies, F. 2001/1887. *Community and civil society*. Harris, J. (ed). Cambridge: Cambridge University Press.

Turkle, S. 2017. *Alone together: Why we expect more from technology and less from each other*. 3rd edn. New York: Basic Books.

Weber, E. 1976. *Peasants into Frenchmen: The modernization of rural France, 1870–1914*. Stanford: Stanford University Press.

Wimmer, A. 2018. *Nation building: Why some countries come together while others fall apart*. Princeton: Princeton University Press.

3 Revisiting the politics of the rural and the Brexit vote

Michael Woods

Introduction

The last decade has seen many liberal democracies shaken by disruptive electoral events that have challenged established political cultures and party systems. The most emblematic of these events were arguably the referendum vote for the United Kingdom to leave the European Union in June 2016 (colloquially known as "Brexit") and the election five months later of insurgent candidate Donald Trump as President of the United States of America, however many countries in Europe and elsewhere have seen growing support for populist and right-wing nationalism parties. Electoral breakthroughs have been achieved, to a greater or lesser degree, for parties including the Rassemblement National (formerly the Front National) in France, Alternative für Deutschland (AfD) in Germany, Vox in Spain, Law and Order (PiS) in Poland, Swedish Democrats in Sweden and others. Outside Europe, high-profile examples include the election of Jair Bolsonaro in Brazil, but also authoritarian-populist turns in countries from India to Turkey.

Whilst the particularities of these disruptive political movements have differed between countries, one commonly observed feature is the apparent significance of a core–periphery or urban–rural distinction in voting patterns. Electoral geography analysis has identified higher levels of support for Brexit in rural and peripheral areas and for Trump in non-metropolitan counties in the USA, as well as patterns of rural support for populist parties more broadly (Los, McCann, Springford, & Thissen, 2017; Scala & Johnson, 2017; Rodríguez-Pose, 2018; Essletzbichler, Disslbacher, & Moser, 2019). Although some dissenting analyses have critiqued this interpretation (Manley, Jones, & Johnston, 2017; Monnat & Brown, 2017), the association between rurality and populism has become embedded in the popular discourse, and the idea of a rural–urban divide has been widely accepted as part of the explanation for disruptive politics. The association is reflected in the proliferation of newspaper headlines such as "Urban–rural splits have become the greater global divider" (*Financial Times*, 31 July 2018), "Like Trump, Europe's populists win big with rural voters" (*New York Times*, 6 December 2016), and "From Trump to Brexit, power has leaked from cities to the countryside" (*The Guardian*, 12 December 2016).

DOI: 10.4324/9781003091714-4

The discursive connection between populism and rural regions has stimulated a resurgence of social science and media interest in the social and economic circumstances of rural communities, with emphasis often placed on perceived economic and political marginalisation (Cramner, 2016; Scala & Johnson, 2017; Rodríguez-Pose, 2018; Wuthnow, 2018; Rodden, 2019; Brooks, 2020; Carolan, 2020; Mamonova & Franquesa, 2020). However, a tendency to respond to disruptive political events as unanticipated shocks and therefore to focus on a search for immediate explanations for these as singular instances frequently misses the longer trajectory of rural discontent that has taken different forms of expression at different times in different places.

Rewind to the start of the 21st century, for example, and we find that a number of countries were in the midst of then largely unprecedented mobilisations of rural activism. In Britain, farmers' protests peaked with blockades of fuel depots in September 2000, whilst a parallel series of mass demonstrations by hunting supporters grew in numbers from 1998 to 2004. In the United States, the emergence of an extremist 'rural militia' movement provoked debate following the Oklahoma City bombing in 1995. In Australia, the populist anti-immigration One Nation party had broken through in state elections in 1998, sparking national discussion of rural discontent. In France, Jose Bové's Confédération Paysanne were mounting anti-globalisation protests, whilst sporadic farm protests also erupted across a number of European countries. The connections between these earlier articulations of rural discontent and the disruptive politics of the 2010s are more apparent in some cases than in others. In France, for instance, rural support for the Front National has oscillated since the 2002 presidential election, whilst in the United States the conservative-populist insurgency in the Republican party can be traced back to origins in rural southern and Mid-West states at the turn of the century. In other cases the lines of continuity are far less clear.

This chapter focuses on the case of Britain and explores possible links between countryside protests in the early 2000s and rural support for Brexit in 2016. At first glance the connections are weak beyond an apparent geographical correspondence. The key issues for the two movements were different (unlike pro-hunting protests elsewhere in Europe, the countryside protests in Britain were not directed toward the EU) and involved different organisations and leaders. However, Brooks (2020) has outlined the potential elements of a trajectory connecting the two periods, centred on "the role of a new political party, the UK Independence Party (UKIP), in occupying a space opened up by the Countryside Alliance, and then vacated by a modernising Conservative Party, as a bridge between the two eras" (p. 791). She observes resonances between the discursive elision of race, rurality and nation in the rhetoric of the countryside protests and the discourses of the Leave campaign in the Brexit debate. As such, the chapter builds on Brooks's exploratory discussion to ask two questions: "Is there evidence of connections between the countryside protests of the early 2000s and the Brexit vote?" and "If yes, in what ways did the countryside protests prepare the ground for rural support for Brexit?"

In order to answer these questions, the chapter revisits data collected in research on the UK countryside protests between 1997 and 2008, predominantly as part of a project on "Grassroots rural protest and political activity in Britain" (2006–2008).[1] Data collected for the project included interviews with 26 activists in rural protest movements and stakeholder organisations, textual analysis of campaign documents and media reports, and a postal questionnaire survey of members of the pro-hunting Countryside Alliance. The survey was undertaken with the cooperation of the Countryside Alliance and involved a paper questionnaire that was posted in May 2007 to 4,344 members in four case study areas: Cheshire, Exmoor and North Devon, Mid Wales and Suffolk. A total of 1,207 useable questionnaires were returned (27.7% response rate). Questions in the survey covered the respondents' participation in rural protests and campaigns, social and economic background and political history, as well as a series of Likert-scale questions on values and political issues.

The politics of the rural and electoral geography

Establishing a connection between recent disruptive politics and earlier rural protests also means establishing a connection with the explanatory frameworks advanced for the countryside protests. In other words, could part of the explanation for outcomes such as Brexit lie in the factors that informed protests by farmers and other rural residents 20 years earlier? In previous work, I argued that the proliferation of countryside protests from the late 1990s represented a shift from a "rural politics" to a new "politics of the rural", in which the focus moved from the distribution and management of resources to the meaning and regulation of rurality itself (Woods, 2003a). From this perspective, the countryside protests marked a break with a long period of relative political and policy stability in rural areas in the global north that had extended from the end of the Second World War. During this period, policies had promoted productivist agriculture, modernisation and the diversification of the rural economy, governed through stable "policy communities" in which rural interests were represented by organisations such as farm unions, working closely with government bodies. In several countries the settlement was reinforced by close links between rural populations, farming unions and centre-right political parties, such as the Conservatives in Britain, Country Party (later the National Party) in Australia, and the Liberal Democratic Party in Japan, which were viewed as the natural representatives of rural voters (Woods, 2015).

The post-war consensus was fractured in the 1980s under the pressure of its internal contradictions. The success of productivist agriculture had diminished fears around food security and majority urban populations were starting to question the financial and environmental cost of "agricultural exceptionalism", whilst the rise of neoliberalism pressed for the deregulation of agricultural markets and open them to global competition. At the same time, the primacy of agricultural interests in rural politics was weakened by the contracting farm population and diversification of the rural economy into areas such as tourism, whose interests sometimes conflicted with those of agriculture. Moreover, in many rural regions, new in-migrants from towns and cities found their pursuit of the "rural idyll" to be compromised by modern farming practices or industrial activity and mobilised to defend their financial and emotional investment. As such, a plethora of local conflicts erupted over issues such as housing development, windfarms, quarrying, new roads, agricultural pollution and nature conservation (Mormont, 1987, 1990; Murdoch & Marsden, 1994; Woods, 1998, 2003a, 2003b, 2005).

The politics of the rural may have started with isolated local protests, but by the late 1990s larger movements were developing, directed towards perceived threats to rural interests from the policies adopted by urban-based governments, such as the dismantling of farm subsidies, trade liberalisation, rationalisation of rural public services, and legislation to regulate or ban traditional activities such as hunting. In these new arenas, control over the discursive representation of the rural was crucial, with emerging rural protest movements working to project their often sectional concerns as broad rural interests ranged against urban interference (Woods, 2003a, 2005, 2008a). Such discursive framing reinforced the perception of an urban–rural divide and helped rural movements to build coalitions of support that could transcend class boundaries and party-political allegiances. Furthermore, the mobilisation of the politics of the rural gave birth to new political actors. With the fracturing of the old political settlement, many rural residents had lost faith in established organisations and parties who were perceived to be no longer able to deliver, and turned to new, often more radical, groups (Woods, Anderson, Guilbert, & Watkin, 2013).

The underlying drivers of the politics of the rural are transnational in scope; however, the timing, form and expression varied between countries, reflecting national differences in context. As argued in Woods (2015), global processes of social and economic change were filtered through national circumstances to manifest as external pressures on established rural political settlements, and combined with internal pressures from change within the countryside and the impacts of national political events, such as changes of government. The effect of these pressures was further shaped by factors such as the nature of the rural policies adopted in the post-war political settlement, the established mechanisms of rural representation and the prevailing discourses of rurality that underpinned the settlement. These factors influenced the timing of expressions of the politics of the rural in specific countries, but also the issues on which mobilisations focused – in some cases agricultural reform, in others nature conservation or anti-hunting legislation, or rural infrastructure and services. Finally, a further set of factors, including patterns of discontent with established representatives, political socialisation, the opportunities presented by the political system, and geography and other structural constraints to political organisation, have all influenced the form that rural political mobilisation has taken – as non-confrontational campaigns, as militant protest movements or as interventions in electoral politics.

In Australia, for example, the dispersed geography of rural areas and pillarised representation of different agricultural sectors militated against the capacity of rural activists to mobilise effective protests or demonstrations. Instead, rural discontent in the late 1990s and early 2000s was channelled into support for populist parties such as One Nation and insurgent local independent candidates, aided by political socialisation that normalised the electoral process as the vehicle for rural representation combined with disaffection with the traditional representative role of the National Party (Woods, 2015). In Britain, by contrast, disaffected rural residents expressed discontent not through voting but through participation in protests and demonstrations, reflecting political socialisation by implicitly copying environmentalist and anti-government protests and feeding into a discourse embellished by the movement leaders that normally placid and apolitical rural people were so angry that they were compelled to protest (Woods, Anderson, Guilbert, & Watkin, 2012).

As such, if the Brexit vote in 2016 did in part reflect an expression of the politics of the rural, it constituted not only a line of continuity from the countryside protests, but also a change in the political environment that pivoted the focus of rural discontent away from protests and demonstrations to the ballot box. The evidence for both these elements is considered in the next section.

The rise and fall of rural protests

From protest to populism in rural Britain?

On 27th July 1997, 120,000 protesters gathered in Hyde Park in London for the Countryside Rally. Organised by the newly formed Countryside Alliance, the rally had been triggered by the landslide Labour victory in the general election in May, and the imminent prospect of new legislation to ban the traditional rural pursuit of hunting foxes and deer with hounds. Whilst defending hunting was the primary motivation for most participants, the Countryside Alliance had calculated that it was too narrow a concern to mobilise a large protest movement and thus had sought to extend the frame by positioning hunting as one of a set of issues including farm incomes, housing and rural services, grouped together under a call to protect the countryside from the policies of an urban-based government. Over the following years, the Countryside Alliance mounted an intensive campaign of demonstrations and lobbies, culminating with the Liberty and Livelihood March in London in September 2002, attended by 408,000 people. More militant spin-out groups, such as the Real Countryside Alliance and the Countryside Action Network, engaged in direct action, including rolling roadblocks. Meanwhile, a separate farmers' protest movement was spawned from impromptu pickets at ports in the winter of 1997/98 – motivated by falling farmgate prices, competition from imports and continuing restrictions on the export of British beef after the BSE epidemic in cattle in the mid-1990s – growing into more widespread spot protests and coalescing around the new Farmers for Action group. The farmers' protests reached a climax in September 2000 with a week of blockades of fuel depots that caused extensive disruption (Woods, 2005, 2015).

The countryside protests generated considerable political heat, aided by substantial coverage in a largely sympathetic media (Woods, 2010a), yet although the catalyst had been an election result, the protests had remarkably little impact on electoral politics. Some fringe parties tried to capitalise on the rural discontent, but with little success. The Eurosceptic UK Independence Party (UKIP) distributed flyers at the Countryside March in 2000 proclaiming that its policies "can help the countryside", including leaving the European Union and the "soil-destroying CAP" [Common Agricultural Policy], whilst the far-right British National Party circulated a *British Countryman* newspaper in rural areas (Woods, 2005). Neither party, however, was rewarded with substantial rural votes. A new Countryside Party similarly failed to gain traction and polled only a handful of votes in the few constituencies that it contested. These dynamics can be explained by the political context of the countryside protests and the framing of the conflict by the Countryside Alliance and the media. Although the legislation to ban hunting was a private member's bill brought by a backbencher and had cross-party support, the blame was clearly attributed to the Labour government. The new leadership of the opposition Conservative Party, meanwhile, conspicuously associated themselves with the countryside protests and promoted the idea that the way to defeat the threat to hunting was to oust the Labour government by voting Conservative. As such, the countryside protests served to reaffirm the historic alignment of rural voters with the Conservative Party (Woods, 2005).

Nonetheless, there was considerable excitement among rural activists that the large numbers attending events such as the 1997 Countryside Rally and 2000 Countryside March represented a significant swing in public opinion against the Labour government that would be reflected at the next election. However, such predictions failed to recognise that the vast majority of countryside protesters were already committed Conservative voters. At the 2001 general election, discontent over hunting, farming and the handling of the recent Foot and Mouth Disease outbreak had only a marginal impact on the Labour vote, even in rural constituencies, with the party losing perhaps only one seat as a result (Woods, 2002). Chastened, rural activists organised themselves for the 2005 election, forming the Vote-OK initiative to direct hunting supporters to help pro-hunting candidates in marginal seats and to encourage tactical voting. The operation achieved greater success, helping to defeat several high-profile anti-hunting MPs, but had little impact on the overall election result (Woods, 2008b).

Prior to the 2001 election, there had been worries in the Labour government about damage to its support from the countryside protests, to which it had responded by seeking a compromise on hunting and staking its own claim to represent the true interests of rural communities with a distinctive policy platform. However, with the realisation that the countryside protests posed it little electoral threat, Labour changed tack after 2001, throwing the weight of the government behind the hunting ban, adopting more radical positions on agricultural policy and losing interest in its own rural agenda (Woods, 2008b). With full government support, the Hunting Act passed in November 2004 and the ban on hunting wild mammals with hounds came in effect in February 2005, confronted by a defiant mass turnout of hunts. Yet, many rural activists were emotionally and physically exhausted and the Countryside Alliance evolved from a protest movement into a more conventional pressure group that forged a working relationship with government, especially after the election of a Conservative–Liberal Democrat coalition in 2010 (Woods et al., 2012, 2013). With this evolution, the links between the rural movement and the Conservative Party were further consolidated. The Chief Executive of the Countryside Alliance, Simon Hart, was elected as a Conservative Member of Parliament in 2010, joining the Cabinet in 2019, and its Deputy Chief Executive John Gardiner was appointed as a Conservative life peer, becoming a minister in 2016. Farm protest leader Brynle Williams, meanwhile, was elected as a Conservative member of the National Assembly for Wales, serving until his death in 2011.[2]

The underlying factors shaping the politics of the rural did not go away, however, and rural discontent continued to foment and bubble around issues of agricultural precarity, windfarm developments, immigration and cuts to rural services, including from the coalition's austerity policies. With the Conservatives now in power and enthusiasm for demonstrations tempered, rural discontent started to find expression in protest votes for UKIP, which had been cultivating a populist platform that opposed wind energy, exploited rural concerns about pressures from Eastern European labour migration, and positioned the EU as the cause of problems in agriculture (Brooks, 2020). The protest votes did not translate into support for UKIP in general elections, but increases in UKIP votes in European Parliament elections, by-elections and local elections persuaded the Conservative leadership that they needed to assuage their Eurosceptic supporters with a pledge in their 2015 election manifesto to hold a referendum on Britain's membership of the European Union.

In the referendum of June 2016, rural districts voted decisively overall to leave the EU, but the detailed electoral geography is more complex. Some of the highest votes for Leave were recorded in parts of East Anglia, Lincolnshire and south-west England that had also supplied large numbers of participants to the countryside protests. Yet, more affluent rural areas such as the Cotswolds – which had been a Countryside Alliance heartland – voted to Remain, as did areas of rural Wales that had been the crucible of farmer protests. Neither was there any clear alignment of leaders of the countryside protests with the Leave campaign. The most prominent link was the Labour MP Kate Hoey, a former Chair of the Countryside Alliance who was a high-profile Leave campaigner. Simon Hart, in contrast, campaigned for Remain, whilst the Countryside Alliance as an organisation adopted a neutral stance in the referendum. Other rural organisations, such as the National Farmers

Union and the Farmers Union of Wales, subsequently lobbied heavily against a "no-deal" Brexit, warning of the potentially severe impact on farming.

Thus, whilst the politics of the rural may have contributed to the political conditions that led to the Brexit referendum, there is no clear continuity between the countryside lobby of the early 2000s and the Leave campaign in 2015. So, why did most rural areas vote for Brexit in 2016? The answer may in part lie with a softer legacy of the countryside protests: their implicit normalisation of populist rhetoric.

Frame continuity: rurality, populism and nationalism

One of the most notable accomplishments of the countryside protests was to convert concerns around a minority interest – hunting with hounds – into a mass movement. This was achieved through the process referred to in social movement theory as "frame extension" (Snow, Rochford, Worden, & Benford, 1986), in which hunting was articulated as an essential component of rural life and a threat to hunting was thereby asserted to be a threat to the rural way of life. As such, the hunting debate was reframed from being a question of animal welfare (as it was framed by hunting opponents) to an issue of rural–urban relations (Woods, 2005), tapping into a deeply engrained cultural trope that constructs the city and the countryside as two separate worlds (Williams, 1973; Woods, 2010b).

The message was reinforced by the branding of the Countryside Alliance and of its major staged events (the Countryside Rally, Countryside March and Liberty and Livelihood March), as well as by the expansion of its platform to cover a wider range of rural issues, including agriculture, housing and access to services. Furthermore, the frame was amplified by explicitly presenting urban political opinion and values as a *threat* to the countryside, as articulated in claims made around the first Countryside Rally in 1997:

> This initiative arose as a response to the frustration and concern felt by country people against the threats posed to the countryside and their jobs, by politicians and urban influence, through prejudice, ignorance and diminishing rural representation.
>
> (Countryside Rally Mission Statement 1997, quoted by Woods 2005, p. 114)

> We cannot and will not stand by in silence and watch our countryside, our communities and way of life destroyed forever by misguided urban political correctness.
>
> (Baroness Mallalieu, Countryside Alliance President, addressing the Countryside Rally 1997, quoted by Woods, 2005, p. 114)

Banners and placards carried by participants in the countryside protests similarly featured slogans such as "Leave the Countryside to Country

People", "Leave Us Alone", and "Say No to the Urban Jackboot". When we surveyed Countryside Alliance members in 2007, 92 per cent of respondents agreed with the statement that "Urban Britain does not care about the countryside", 97 per cent agreed that "Rural culture is under threat" and 85 per cent agreed that "Rural people are an oppressed minority" (Table 3.1). As the countryside protests failed to realise their objective of stopping the ban on hunting with hounds, the rhetoric of an urban polity out of touch with rural people became a self-perpetuating driver of rural discontent.

The idea of a rural–urban divide had been present in British politics before the countryside protests and had implicitly underpinned the Conservative hegemony in rural areas (Woods, 2005), but it had been weakened by the decline of agricultural employment, migration and the blurring of rural and urban cultures, economies and social structures. The countryside protest movement brought it back to the forefront of national politics, such that it was available as a convenient hook on which to hang spatial imaginaries mobilised during and after the 2016 Brexit referendum that pitted "Left Behind Britain" against a self-serving metropolitan elite (Sykes, 2018) (and via a further frame extension to encompass lagging ex-industrial regions and provincial towns in the "Left Behind Britain" category). The mischievous implication in the discourse that EU membership only benefitted the metropolitan centres of Britain resonated with rural and provincial voters convinced of a structural bias in British politics, in spite of factual evidence to the contrary, thus leading to strong votes to leave the EU in rural and ex-industrial districts that were most heavily dependent on EU markets and received the greatest proportion of EU structural funding (Los et al., 2017).

The discursive frame of the countryside protests was further amplified by associating rural identity with national identity, again drawing on a deeply embedded cultural narrative that had been periodically evoked by Conservative political leaders from Stanley Baldwin to John Major (Woods, 2005). Revived by the countryside protest movement, the discourse became an accusation that proposals to ban hunting were somehow un-British and

Table 3.1 Perceptions of members of the Countryside Alliance on rural–urban relations, 2007

	Strongly agree	Agree	No strong opinion	Disagree	Strongly disagree	n
Urban Britain does not care about the countryside	54.3%	37.3%	5.7%	2.6%	0.1%	1197
Rural culture is under threat	65.0%	32.0%	2.0%	1.0%	0.0%	1201
Rural people are an oppressed minority	52.3%	32.9%	8.9%	5.5%	0.3%	1178

Source: Survey of Countryside Alliance Members, 2007: http://doi.org/10.5255/UKDA-SN-6238-1.

that the loss of rural tradition would erode distinctive British culture. As a leader comment in *The Daily Telegraph* newspaper put it before the Countryside March in 1998:

> The people who are coming to London are the backbone of the nation. They are those who have always been ready to fight for their country when required. For them "country", in the sense of nation, is closely bound up with "country" in the sense of green fields.
>
> (Leader article in *The Daily Telegraph*, 28 February 1998, quoted by Woods, 2005, p. 116)

The notion of rallying rural people to defend British values was played out in the dramaturgy of the countryside protests, with many participants carrying Union Jacks, St George's cross flags of England, and the national flags of Scotland and Wales. A limited licence radio station for the Countryside March played stirring patriotic music that suggested to one journalist that "you can only be truly British if you live in the countryside and like to kill animals" (*The Guardian*, 2 March 1998, quoted by Woods 2005, p. 117). In the later stages of the protests, the militant breakaway Real Countryside Alliance group fly-posted rural road signs with stickers showing a green Union Jack.

For the countryside protesters the threat to British national identity came from within, from urban voters and political leaders whose loss of rural roots had eroded their sense of British values and opened their minds to polluting foreign and cosmopolitan influences. However, with this idea seeded in prolific media commentary on the countryside protests, it was not too difficult for attention to pivot to the perceived external threat to British identity from the European Union, again reinforced by one-sided reporting by Eurosceptic newspapers. The elision of EU membership and debates over immigration, with xenophobic references to not hearing English on the London underground, for example, further resonated with prejudices that the capital was no longer truly British and fears that rural areas could follow. This was arguably reflected in strong votes for Leave in rural districts that had received disproportionately high inward migration from Eastern Europe after 2004, such as Lincolnshire, and where local conflicts had developed around competition for jobs and pressure on public services (Brooks, 2020).

Reflecting back on the countryside protests from a post-Brexit standpoint, it is the populism of the countryside protests that really stands out. Political figures in Britain from both sides of the partisan divide had employed populist rhetoric on occasion, but the countryside protests were arguably the first overtly populist mainstream political movement in Britain for nearly a century. Within the discursive framings of the rural–urban divide and the defence of British values was the simple idea that the protests represented "the people" coming together to take on a distant and undemocratic "elite". Baroness Mallalieu's address to the Countryside Rally in 1997 included the declaration that "the countryside has come to London to speak out for freedom" (quoted by Woods, 2005, p. 115), whilst a banner carried by long-distance marchers to the same event (mis-)quoted lines from G. K. Chesterton: "Do not quite forget; We are the people of England, who have not spoken yet".

The populist framing of the countryside protests was an important tactical device that enabled organisers to overcome the inherent conservatism of their grassroots supporters. Only one in ten Countryside Alliance members surveyed had participated in a protest or demonstration before 1997, and the traditional political socialisation of the countryside had suggested that radical politics, strikes and protests were urban and alien (Woods, 2005). The populist rhetoric allowed organisers to distinguish the countryside protest from left-wing demonstrations and reassured participants by positioning them in a long history of popular uprisings that had created the British constitution, extending back to the mediaeval Peasants' Revolt. This message was repeatedly emphasised in speeches and sympathetic press columns, and in an extensive feature in the commemorative magazine for the 2002 Liberty and Livelihood March.

As such, the countryside protests helped to normalise populist discourse in Britain, facilitating the adoption of similar language by pro-Brexit campaigners a decade later, in the vilifying of 'elitist' pro-EU sentiments in the referendum campaign and in the subsequent discursive pitching of 'the people' against Parliament and the Supreme Court in struggles over Brexit legislation.

Trust, discontent and radicalisation

The line of continuity from the countryside protests to Brexit is constituted by resonances in the frames and discourses mobilised rather than through organisational ties, but there is a further similarity. In each case the rhetoric of populism was duplicitous, with both the countryside protests and Brexit arguably being elite-led populist movements. Although the countryside lobby was keen to counter representations of hunting as an upper-class pursuit, efforts were undermined by the significance of large landowners and aristocrats in the leadership tiers of the Countryside Alliance and in funding its activities (RPM, 2002). Similarly, the populism of the Leave campaign belied the financial backing of millionaire business owners and the privileged, privately educated background of many of its prominent advocates. The apparent paradox may, in part, be explained by the particularities of the populist discourses mobilised.

In both the countryside movement and the Brexit campaign, "the people" were discursively pitted quite specifically against a *metropolitan* elite, comprising "professional politicians", lawyers, journalists, "experts" and the so-called "chattering classes". This framing struck out against a professionalised political culture that was centred on London and therefore perceived to be out of touch with people in rural areas and provincial towns. Significantly, this framing also resonated with the disillusionment of rural people with the (professional) organisations that had represented their interests in the post-war political settlement that is a key assertion of the "politics of the rural"

Table 3.2 Perceptions of members of the Countryside Alliance on political institutions, representation and radicalism, 2007

	Strongly agree	Agree	No strong opinion	Disagree	Strongly disagree	N
The Conservatives no longer represent rural interests	8.6%	37.4%	27.4%	23.5%	3.1%	1174
I have become more militant than I was 10 years ago	25.3%	37.0%	21.2%	14.6%	1.9%	1160
Britain should leave the European Union	32.3%	18.9%	25.2%	18.0%	5.7%	1165

Source: Survey of Countryside Alliance Members, 2007: http://doi.org/10.5255/UKDA-SN-6238-1.

(Woods, 2003a; Woods et al., 2013). Yet, in losing trust in elected politicians and pressure groups such as farming unions, it is not inconceivable that disaffected rural residents might have turned back to older, paternalistic leaders – people who the traditional political socialisation of rural society had presented as disinterested public servants working for vertically integrated rural communities (Woods, 2005).

The Conservative Party itself was not immune to disaffection. Even though just over half of the Countryside Alliance members surveyed in 2007 were also members of the Conservative Party, 46 per cent agreed with the statement that 'The Conservatives no longer represent rural interests' (Table 3.2). A primary concern with hunting and farming meant voting Conservative in general elections to block the Labour Party, but with a Conservative government in place after 2010 the weakened partisan loyalty allowed continuing rural discontent to find expression in local and European elections and parliamentary by-elections in protest votes for the Eurosceptic UKIP and Brexit Party. Indeed, the countryside protests emboldened rural activists to diverge from their political socialisation, with 63 per cent of survey respondents stating that they had become "more militant than I was ten years ago".

Given the distrust of mainstream politics indicated by these figures, it is not surprising that the survey also found that 51 per cent of respondents thought that Britain should leave the UK, compared to only 24 per cent wishing to remain. In other words, in 2007 the grassroots membership of the Countryside Alliance was already considerably more Eurosceptic than either general public opinion at the time or the leadership of the rural movement. It was this underlying conviction that carried through into rural support for Brexit in the 2016 referendum.

Conclusion

There are strong resonances between the countryside protests in Britain in the late 1990s and early 2000s and rural support for Brexit in 2016, yet the actual relationship between the two is complex. Rural communities were divided in the EU referendum, just as the country as a whole was. There is evidence that the farming vote was split 50/50 between Leave and Remain (Brooks, 2020), and a number of rural areas voted to stay in EU, especially in south-east England, Scotland and parts of Wales. Organisations active in the countryside protests were generally neutral on Brexit, and several prominent protest leaders campaigned to remain. At the grassroots level, however, many rural residents who participated in the countryside protests are likely to have also voted to leave the EU.

As Brooks (2020) has observed, the countryside protests helped to create the conditions for the EU referendum by generating volatility in Conservative support that was exploited by UKIP. Moreover, the countryside protests established a legacy in normalising populist discourses that were later reprised in the Brexit campaign, notably the vilification of a cosmopolitan metropolitan elite. Ultimately, however, it would be more accurate to say that the countryside protests did not lead to Brexit, but that both were articulations of an underlying 'politics of the rural'. Both movements tapped into rural discontent with perceived economic and political marginalisation and fears over threats to a traditional way of life. At the turn of the century, the political context channelled this discontent into demonstrations and protests. A decade and a half later, the exceptional circumstance of the EU referendum created a unique opportunity for rural discontent to be expressed again, contributing to a coalition of interests that delivered the Leave result.

Notes

1 "Grassroots Rural Protest and Political Activity in Britain", funded by the Economic and Social Research Council (ESRC) (Award RES-000-23-1317), 2006–2008. I am grateful to co-investigator Jon Anderson and research assistants Steven Guilbert and Suzie Watkin for their work on this project.
2 Original research by author, with information sourced from news reports and official listings online of UK government ministers and members of the House of Commons, House of Lords and National Assembly for Wales.

References

Brooks, S., 2020. Brexit and the politics of the rural. *Sociologia Ruralis*, 60(4), pp. 790–809.

Carolan, M. S., 2020. The rural problem: Justice in the countryside. *Rural Sociology*, 85, pp. 22–56.

Cramner, K. J., 2016. *The politics of resentment*. Chicago: University of Chicago Press.

Essletzbichler, J., Disslbacher, F. and Moser, M., 2019. The victims of neoliberal globalisation and the rise of the populist vote: A comparative analysis of three recent electoral decisions. *Cambridge Journal of Regions, Economy and Society*, 11, pp. 73–94.

Los, B., McCann, P., Springford, J. and Thissen, M., 2017. The mismatch between local voting and the local economic consequences of Brexit. *Regional Studies*, 51(5), pp. 786–799.

Mamonova, N. and Franquesa, J., 2020. Populism, neoliberalism and agrarian movements in Europe: Understanding rural support for right-wing politics and looking for progressive solutions. *Sociologia Ruralis*, 60(4), pp. 710–731.

Manley, D., Jones, K. and Johnston, R., 2017. The geography of Brexit – What geography? Modelling and predicting the outcome across 380 local authorities. *Local Economy*, 32(3), pp. 183–203.

Monnat, S. M. and Brown, D. L., 2017. More than a rural revolt: Landscapes of despair and the 2016 Presidential election. *Journal of Rural Studies*, 55, pp. 227–236.

Mormont, M., 1987. The emergence of rural struggles and their ideological effects. *International Journal of Urban and Regional Research*, 7, pp. 559–578.

Mormont, M., 1990. "What is rural?" or "how to be rural": Towards a sociology of the rural. In Marsden, T., Lowe, P. and Whatmore, S. (eds.) *Rural restructuring*. London: David Fulton, pp. 21–44.

Murdoch, J. and Marsden, T., 1994. *Reconstituting rurality*. London: UCL Press.

Rodden, J. A., 2019. *Why cities lose*. New York: Basic Books.

Rodríguez-Pose, A., 2018. The revenge of the places that don't matter (and what to do about it). *Cambridge Journal of Regions, Economy and Society*, 11, pp. 189–209.

RPM, 2002. *The rich at play: Foxhunting, land ownership and the "countryside alliance"*. London: Revolutions Per Minute.

Scala, D. J. and Johnson, K. M., 2017. Political polarization along the rural-urban continuum? The geography of the Presidential vote, 2000–2016. *Annals of the American Academy of Political and Social Sciences*, 672, pp. 162–184.

Snow, D. A., Rochford, B. E., Worden, S. and Benford, R., 1986. Frame alignment processes. Micromobilization and movement participation. *American Sociological Review*, 51, pp. 464–481.

Sykes, O., 2018. Post-geography worlds, new dominions, left behind regions, and "other" places: Unpacking some spatial imaginaries of the UK's Brexit debate. *Space and Polity*, 22, pp. 137–161.

Williams, R., 1973. *The country and the city*. Oxford: Oxford University Press.

Woods, M., 1998. Researching rural conflicts: Hunting, local politics and actor-networks. *Journal of Rural Studies*, 14, pp. 321–340.

Woods, M., 2002. Was there a rural rebellion? Labour and the countryside vote in the 2001 general election. In Bennie, L., Rallings, C., Tonge, J. and Webb, P. (eds.) *British elections and parties review, volume 12 – The 2001 general election*. London: Frank Cass, pp. 206–228.

Woods, M., 2003a. Deconstructing rural protest: The emergence of a new social movement. *Journal of Rural Studies*, 19, pp. 309–325.

Woods, M., 2003b. Conflicting environmental visions of the rural: Windfarm development in Mid Wales. *Sociologia Ruralis*, 43, pp. 271–288.

Woods, M., 2005. *Contesting rurality*. Aldershot: Ashgate.

Woods, M., 2008a. Social movements and rural politics. *Journal of Rural Studies*, 24, pp. 129–137.

Woods, M. (ed.), 2008b. *New Labour's countryside*. Bristol: Policy Press.

Woods, M. 2010a. Reporting an unsettled countryside: The news media and rural protests in Britain. *Culture Unbound*, 2, pp. 215–239.

Woods, M., 2010b. *Rural*. London: Routledge.

Woods, M., 2015. Explaining rural protest: A comparative analysis. In Strijker, D., Voerman, G. and Terluin, I. J. (eds.) *Rural protest groups and populist political parties*. Wageningen: Wageningen Academic Publishers, pp. 35–62.

Woods, M., Anderson, J., Guilbert, S. and Watkin, S., 2012. 'The country(side) is angry': Emotion and explanation in protest mobilization. *Social and Cultural Geography*, 13, pp. 567–587.

Woods, M., Anderson, J., Guilbert, S. and Watkin, S., 2013. Rhizomic radicalism and arborescent advocacy: A Deleuzo-guattarian reading of rural protest. *Environment and Planning D: Society and Space*, 31, pp. 434–450.

Wuthnow, R., 2018. *The left behind*. Princeton: Princeton University Press.

4 Populism of the dispossessed

Rethinking the link between rural authenticity and populism in the context of neoliberal regional governance

Bianka Plüschke-Altof and Aet Annist

4.1 Introduction

This chapter sets out to rethink the link between rural authenticity and populism. While it acknowledges the way in which populist movements employ rural authenticity by presenting rurality as home to true and unspoilt identities, it questions this default connection. Instead, we argue that the rise of populist sentiments may be the result of a turn towards rural authenticity as a way of enhancing regional development in the context of an increasing neoliberalisation of regional governance. In fact, the example of a heritage culture regime in the southern Estonian Seto region[1] demonstrates that how rural authenticity is produced and used as a regional development strategy might result in benefits for rural populations, but at the same time produces new exclusions and dispossessions among them. The exclusion from the boundaries of a narrowly defined rural authenticity might then create fruitful ground for a drift towards populist sentiments by those facing dispossession. Considering the peripheralisation of the Central-Eastern European (CEE) countryside since the post-socialist transformation (Kay, Shubin, & Thelen, 2012; Lang et al., 2021), neoliberal regional policies are thus discussed as the often overlooked link between rural authenticity and populism.

The theoretical background connects debates on rural authenticity with the concept of dispossession in neoliberal governance regimes to critically discuss who benefits from producing images of rural authenticity and the institutionalisation of heritage culture, and who does not. Methodologically, we bring together the results of repeated fieldwork in Setomaa since 2002. While originally focusing on heritage culture and social and symbolic dispossessions (Annist, 2013, 2015, 2017) as well as rural peripheralisation and stigmatisation (Plüschke-Altof, 2018; Plüschke-Altof & Grootens, 2019; Plüschke-Altof, Loewen, & Leetmaa, 2020), this data will now be critically scrutinised from the perspective of rural authenticity production (see also Kuutma and Annist 2020). Based on this framework, we answer questions on (1) how authentic rurality is (re)produced in Setomaa against the backdrop of post-socialist neoliberalisation of regional politics, and (2) the way in which rural authenticity as celebrated best-practice strategy for regional

DOI: 10.4324/9781003091714-5

development exacerbates dispossessions of the rural population, which might result in attraction towards populist sentiments.

4.2 Populism of the dispossessed and neoliberal use of rural authenticity: conceptualisation

Authenticity and heritage culture are subjects researched in many disciplines. Former studies on rural authenticity focus on the production of rural idyllic images, the socio-spatial context enabling their persistence and marginalisation resulting from the contradiction between such idyllic images and realities of lived space (for an overview, see Baumann, 2018). As discussed by Baumann (2018) in the case of Germany, idyllic representations of the rural are often linked to nationalist sentiments. In the form of escapism, the rural idyll has been positioned against urban stereotypes during the urbanisation process since the industrial revolution. Throughout the rise of nationalism in the 18th century, the rural idyll was connected to the idea of the countryside as "Heimat" (*homeland, cradle of the nation*), which today manifests in regionalism as a counterweight to globalisation that continues to be employed by populist movements (ibid.).

However, in this chapter we argue that in CEE in particular, it is not necessarily the targeted employment of idyllic rural images by populist movements that links rural authenticity to populism. Instead, the attraction to populism in regions represented as rurally authentic might result from the entanglement of rural authenticity regimes with capitalist spatialities. As demonstrated by Halfacree (2006, 2007), the production of rural authenticity and rural idyll can be critically positioned within the workings of capitalism. In the form of a "consumerist idyll", idealised representations of the rural convey a shift from agricultural productivism towards practices of spatial consumption within post-productivism wherein rural landscapes, aesthetics and lifestyles are treated as endogenous resources for "place-selling" (Bristow, 2005) via marketing and branding (Ashworth, 2009). Instead of agricultural production, regional development is thereby related to consumption-oriented practices of leisure, counter-urbanisation and second-home dwelling (Halfacree, 2007).

Following this line of argumentation, the turn towards rural authenticity can be seen as a profitable and – in neoliberal regional policy regimes – also a *necessary* development strategy. In CEE in general (PoSCoPP, 2015) and Estonia in particular (Loewen, 2018), the neoliberalisation of regional policy coincided with post-socialist transformation from a planned to a market economy. For rural areas, this manifested in tangible rural peripheralisation – objectively as well as subjectively – that is particularly pronounced in the Baltic states (Lang et al. 2021). This meant a gradual migration of jobs, people and services to larger towns and cities. At the same time, the countryside and its inhabitants were subjected to territorial stigmatisation as the "downside" of the centre–periphery, urban–rural and east–west divides (Kay et al., 2012; Plüschke-Altof, 2016) and gradually lost their role in nation-making, experiencing a type of symbolic dispossession (Annist, 2017). Most importantly for our case, the neoliberalisation of regional governance also entails a shift of responsibility for dealing with regional development: from redistributive state-level policy to policy based on regional competitiveness and the use of endogenous resources (Bristow, 2005). Next to the responsibilisation of regions for enforcing local development irrespective of their different resources and capacities to do so (Loewen, 2018), the responsibility for a self-inflicted "failure" to participate in the fruits of neoliberal development has also shifted (Annist, 2005; Plüschke-Altof & Grootens, 2019).

Against this backdrop, the "countryside industry", including rural authenticity and heritage culture, eco-lifestyles, organic farming and regional product labelling, has become a widely recognised development strategy (Halfacree, 2006; Woods, 2013; Kašková & Chromý, 2014). The preconditions for following a development path based on rural authenticity are, however, not equal. A region may already be recognised for authentic heritage culture, or it may be teased out under the pressure for endogenous development and the diversification of the rural economy. Regions with a recognised heritage culture, such as the Estonian region of Setomaa, with its authenticity regime based on "rural idyll with living folklore" (Palang, Semm, & Verstraete, 2009), have a considerable advantage in these conditions.

Earlier studies emphasise the empowerment brought to the locals who have taken control of their own authenticity creation and heritage management (Sahlins, 1999; Bunten, 2004, 2008). Whilst such research places the two parties involved in the quest for authenticity – the tourist and the locals – on an equal footing, and even declares heritage inalienable by the tourists in that tourist presence, gaze or consumption does not take away the value of heritage for the locals, the problematisation of such processes is more recent. One point of critique has been the commodification of rural authenticity, where studies warn of place commodification "hijacking culture" (Kavaratzis & Ashworth, 2015) so that rural areas turn into "theme parks for the urban middle classes" (Fairly 1999 in Halfacree 2006: 329) and become "locked in the past" (Annist 2004, as qtd. in Palang et al., 2009: 101). With the arrival of critical heritage studies (Smith, 2006; Bendix, 2009; Harrison, 2013), the political nature of heritage has been subjected to scrutiny (Winter, 2015; Kuutma, 2019), conveying how heritage cultures are not merely empowered marginals but may generate new heritage elites (Escallon, 2017) and heritage hegemonies (Annist, 2013, 2018). The results of such formations are studied considerably less. The studies that do address this, frame it in terms of the dispossession amongst the population not benefitting (Annist, 2018; Hayes, 2020), or focus on the resulting exclusions (Annist, 2013; Storie, Chenault, Külvik, & Bell, 2020).

We will use the concept of dispossession to make sense of the losses experienced by the part of the Seto population aware of the creation of new heritage and authenticity regimes, yet unable to benefit, and will demonstrate that this plays a part in becoming attracted to populist parties. The concept of dispossession is borrowed from Harvey (2004, 2007) whose interpretation of neoliberalism as a political economy of capitalist expansion led by the interests of a global entrepreneurial class has chronicled how neoliberal policies became global. He understands neoliberalism as a regime of "accumulation by dispossession" (Harvey, 2004) and an aggressive process of commodification, privatisation and redistribution of resources. In the "classic" process of dispossession, people are dispossessed of their means of production, subjugating them to pressures from the capitalists acquiring those means and transferring wealth from public to private ownership. Over the years, the concept has been applied to analyse a variety of losses that the disenfranchised suffer: losses in the cultural (Kalb, 2009a; Creed, 2011), moral (Hann, 2011) and economic sphere (Kasmir & Carbonella, 2008) experienced mostly by deprived groups of the society, often spiralling into further losses. In neoliberal contexts, dispossession and the development of new class structures are strongly interlinked (Harvey, 2004). Annist (2015, 2018, 2022) combines the terms "social dispossession" and "symbolic dispossession" to understand how social fragmentation, particularly in regions suffering from a loss of status in the neoliberal competitive settings, is simultaneously caused by neoliberalism and then furthers its causes, making these concepts well suited to understand how the aims of authenticity creation shift local relations and realities.

As recent studies indicate (Kalb, 2009a, 2009b; Pilkington, 2016; Narotzki, 2019), dispossessions contribute to the appeal of populist parties. Indeed, populism is carried by a sense of dispossession, as the dominant sociopolitical systems appear to have a preference for or discursive attention to interests that do not serve the "common people" (see Mazzarella, 2019). This is not necessarily what it has always referred to, however. There has been a shift in what is understood as populism in the public arena – a shift that is clearer, perhaps, in the East European contexts. In the 1990s, populism referred mostly to attempts at relaxing social policies, thus straying from neoliberal fiscal regimes. The populism that has emerged since 2015 has a rather different hue, and does not concentrate on economic liberalism and fiscal policies. Indeed, the role these might have played in the worsening conditions and disenfranchisement of some social groups and the peripheralisation of rural areas often remains unquestioned by those adopting populist rhetoric. Instead, the focus has shifted to opposing social liberalism, in particular to rejecting any consideration given to a variety of minorities. This new focus is particularly interesting within the hegemony of Seto heritage, where the "cultural elite" has directly benefited from the minority status of Seto people.

4.3 Methodological approach and database

The case study is based on extensive research in Setomaa since 2002 which employed anthropological long-term fieldwork and focused ethnography, including participant observation and in-depth interviews with representatives from the local heritage culture coalition such as politicians, administrators, cultural workers, entrepreneurs and representatives from the marketing

Table 4.1 List of cited interview partners

No	Name	Gender	Field
1	Andres	M	Culture
2	Anne	F	Summer resident, not Seto
3	Diana	F	Social work
4	Erki, Jüri, Tarmo	M	Residents with self-declared Seto and Estonian origin
5	Greeta	F	Journalism, culture
6	Jaagup	M	Entrepreneurship, community initiative
7	Kaie	F	Resident and self-declared Seto
8	Karl	M	Politics
9	Marianna	F	Marketing, culture
10	Marko	M	Entrepreneurship, culture
11	Mart	M	Resident and self-declared Seto
12	Märt	M	Politics, culture
13	Peeter	M	Resident with some Seto roots, self-declared Estonian
14	Ragnar	M	Entrepreneurship, culture
15	Stiina	F	Social work
16	Tõnis	M	Politics, culture
17	Toomas	M	Culture, politics

and tourism sectors. However, we were also carefully searching for the voices of those not part of the heritage culture coalition. The sampling was based on initial contacts identified with the help of a context analysis that scrutinised former studies on Setomaa in particular and rural development and the Estonian heritage culture movement in general, development programmes and socio-economic data. Altogether, in the course of this long-term engagement in the field, more than 100 interviews were conducted. To preserve the anonymity of the interviewees in a very sparsely populated area, we use pseudonyms (Table 4.1) and only indicate their role/field of engagement, with their position or the institutions they represent not specified.

4.4 The prime example of radical neoliberalisation: case study introduction

The turn towards rural authenticity and heritage culture in Setomaa took place against the backdrop of an increasing neoliberalisation of (regional) policy in Estonia. After regaining independence in 1991, neoliberal economic reforms in the form of "shock therapy" were established as a priority and enforced with a rigidity unlike in any other post-socialist country (Lauristin & Vihalemm, 2009; Bohle & Greskovits, 2012). Estonia is thus often considered a "prime example of radical, consistent and successful market-economy reforms in Central and Eastern Europe" (Wrobel, 2013: 17). Next to the tremendous social impacts of Estonia's radical neoliberalisation process demonstrated in former studies (Lauristin & Vihalemm, 2009; Pungas, 2017), it also affected spatialities. Following the path of EU cohesion policy, regional policy started to focus on regional competitiveness and economic growth, gradually taking the shape of consumption-oriented place promotion and post-productivist entrepreneurialism (Loewen, 2018; Plüschke-Altof et al., 2020). While this offered new opportunities for the diversification of the rural economy (for example, rural tourism), these new possibilities could not compensate for the rapid drop of the population involved in agriculture from 25% in 1989 to 7% in 2003 (Vöörmann, 2005: 75). Instead, the regional divide between "Tallinn and the rest" (Plüschke-Altof et al., 2020) increased, manifesting in a rapid (sub-)urbanisation while at the same time peripheralisation in small towns and the countryside deepened. This divide also manifests in public discourse which tends to equate the rural with the peripheral by portraying rural areas as lagging behind economically, socially problematic, politically dependent and institutionally thin (Plüschke-Altof, 2016), as well as responsible for their inability to participate in economic growth generated by the neoliberal development path (Plüschke-Altof, 2017). Such simultaneous peripheralisation and stigmatisation has been confirmed for other rural areas in CEE as well (Kay et al., 2012; PoSCoPP, 2015). However, rural areas are also represented with idyllic imagery (Plüschke-Altof et al., 2020), especially in rural tourism, marketing, and heritage regions that can build on a national identity discourse constructing Estonians as country people (Nugin, 2014).

Rural peripheralisation was, and is still, also prevalent in Setomaa. Situated on the border with Russia, the Seto region was incorporated into the Estonian state in 1920 as the historical region of Pechory (*Petserimaa*). After the Soviet occupation of Estonia during World War II, about three-quarters of Pechory was unified with the Pskov oblast of the Russian Soviet Federative Socialist Republic, leaving only a quarter of it within the territory of today's Estonian Republic. After Estonia regained its independence in 1991 and joined the Schengen Area in 2004, this resulted in a division of Setomaa demarcated by the external border of the EU. This meant loss of land and property that people had maintained across the mostly symbolic border during the Soviet era, and new complications for access to the graves of their ancestors. Further, access to the cross-border farmers' market, a substantial source of income, also disappeared. The former system of state and collective farms (*sovkhozes* and *kolkhozes*) broke down, leading to loss of employment in agriculture from 25% to 7%, or 130,000 people (Vöörmann, 2005: 74–76), the worst loss percentage amongst all ex-Soviet countries. The cut off from former trading routes, the restructuring of the rural economy and ongoing urbanisation caused a decrease in agricultural and associated economic activities, unleashing a demographic shrinkage process spurred by outward migration. Today, Setomaa, one of the most peripheral regions, can be described as a sparsely populated rural area (3,280 inhabitants;[2] population density 7.1 per/km²) with relatively high deprivation (10% in 2019, 4th position in Estonia), lower than average income levels (80% of Estonian average), and relatively high emigration levels. Furthermore, due to the distinctiveness of the Seto region (local dialect, cultural differences, Orthodox religious history separating the area from Estonian Lutheranism), they have long been the internal Other (Lõuna, 2003; Valk & Särg, 2015). However, this, along with the painful separation of Setomaa since the 1990s, has carved a development path which has increasingly celebrated heritage culture and rural authenticity and has become a best-practice example for the employment of endogenous resources (Raagmaa, Masso, Reidolf, & Servinski, 2012).

4.5 Setomaa "Yours Authentically": the institutionalisation of a rural authenticity regime

The Seto region has risen to prominence since the 1990s, first due to its protests against the dilution of the Tartu Peace Treaty, the legal cornerstone of the Estonian Republic which was originally rejected unilaterally by the Russian Federation, leading to the aforementioned separation. But as these efforts failed and Setomaa remained divided between Russia and Estonia, this original motivator has given rise to a variety of well-established institutional and funding structures, as well as an active cultural, political, and economic elite. On this basis, the Seto have successfully established their local identity and their region's cultural uniqueness, which has fostered a particular kind of local development.

The rural authenticity regime builds on an image-reversal (Plüschke-Altof, 2018) that turns existing images of regional peripherality and cultural peculiarity on their head so that the peripheral location becomes the exact reason why Setomaa has "remained special" (Marianna, marketing), "unique and genuine" (Külvik, 2014). The border location becomes the "gateway to Europe" (Karl, politics) and the region is described in a pastoral narrative of "pure nature" and "fresh air" (Greeta, media and cultural worker) juxtaposed with urban stereotypes such as "car theft, drug addiction, traffic jams" (Tõnis, politics and culture). This image-reversal put forward by the Seto elite works as a resurrection narrative, wherein alleged deficits of the region are restated and shifted to the past while portraying the present as a success story, and the base of a new sense of pride among Seto people. The Seto elite has thereby reinvented Setos as pioneers of a post-productivist development path based on heritage culture and entrepreneurship.

This new image is propagated via different marketing and journalistic channels, such as the Setomaa portal in the national daily *Postimees* and various regional websites. One of the main outreach events drawing attention to Seto heritage culture and the ability to organise popular events is the Seto Kingdom Day *(Seto Kuningriigipäiv)*, which has taken place since 1994. It has gradually developed from a locally oriented into a touristic event displaying popular features of Seto heritage culture, such as handicraft, folk costumes, local cuisine and traditional Leelo choir singing, which is on the UNESCO Representative List of Intangible Cultural Heritage. Each summer, thousands of tourists and locals gather in Setomaa to celebrate the Kingdom and partake in the election of the "Seto king" *(ülembsootska)*, who is supported by a council of predecessors *(Kroonikogo)*. In addition to the Kingdom Day, several marketing campaigns, such as Leelo Day *(Leelopäiv)*, Pop-Up Café Day *(Kostipäiv)* and the cultural-touristic road *(Seto Külavüü)*, disseminate the image of living history in Setomaa. United by the slogan "Yours authentically" (in Seto, *kimmäs kotus*, or strong, but also authentic home), they all draw on existing ascriptions of Seto culture as being exotic and authentic.

An institutionalisation of this rural authenticity regime was enabled by the (re)foundation of two organisations: The Seto Congress (*Seto Kongress*) and the Union of Setomaa (*Setomaa Liit*). While the former functions as the representative body of Setomaa and the Seto people, the latter coordinates the political and developmental activities of the region. The Congress meets every three years to make decisions concerning the cultural, economic and political development of the region. Relevant actions are then implemented by an elected Council of Elders (*Vanõbide Kogo*) and the Union of Setomaa, consisting of representatives from the local administration and an executive body. A crucial role is played by a monthly roundtable (*tsõõriklaud*) initiated by the Union and coordinating the activities of the main actors, including a variety of umbrella organisations of tourism, handicraft, arts and theatre, entrepreneurship and renewable energy. The Union also acts as a lobbying organisation with close ties to the Setomaa support group in the Estonian parliament (*Setomaa Toetusrühm*).

Many of those structures were established and maintained by Seto enthusiasts – or, as the cultural worker Andres calls it, a "handful of crazy people", lobbying for investments, state support and funding by emphasising the important role of Setomaa in (1) securing the borders of Estonia and the EU, (2) preserving the UNESCO heritage culture and (3) sustaining diversity in times of globalisation and homogenisation when "every day we have one less language" (Marianna, marketing). When justifying, for instance, why local schools should be preserved, they argue "through culture" as Märt (politics, culture), actively engaged in lobbying work, explains. For Marko, active in the field of entrepreneurship and culture, it is "the entering [of Leelo] in UNESCO that has given us here locally this kind of trump card so that we can always use it as an argument". This institutional layer supports Setomaa's regional development, but it is also further sustained by the state-funded Setomaa Cultural Programme and Setomaa Development Programme – today united into the Setomaa Programme – which has supported a variety of projects, from cultural to entrepreneurial endeavours, deemed to develop the Seto region.

4.5.1 Rural authenticity as an answer to neoliberal calls

As indicated by such state funding attracted to Setomaa, the use of rural authenticity has proven quite promising in the context of regional competitiveness. But it also creates several dilemmas. As Bristow (2005) has pointed out, instead of a distribution of resources on the premise of neediness, regional competitiveness ultimately results in a zero-sum logic wherein not everyone can be a winner. While Setomaa has attracted tangible resources, other regions have not been as "successful" in fundraising. This dominance of Setomaa in comparison to other regions is critically scrutinised from the outside. Most of the interviewees are aware that this criticism reflects a certain political favouritism towards Setomaa, but interpret it as a form of "jealousy about certain things, above all the support programme" (Marko, entrepreneurship and culture). While this local reading mirrors the existing advantages the region enjoys, it similarly downplays their regional policy impacts in times of resource scarcity. Similarly, the uneven distribution of benefits from the chosen authenticity-based development path within Setomaa is explained by the missing "orientation towards success" (Marianna, marketing) or inherent "weakness" (Greeta, journalism and culture) of the respective localities.

Moreover, building rural development on authenticity entails the risk of a neoliberal co-optation. The central aim of these efforts has been entirely justified: to create a positive image that attracts investors who previously saw "no sense in investing here" if there is only "depression" and "lack of belief in oneself", as Toomas (culture and politics) remembered from the 1990s. This also includes "putting Setomaa in the picture" or "helping our helpers" with public celebrations of successful coping efforts and cultural uniqueness, as Toomas and Tõnis recall from their activities in the field of culture and politics. Greeta, a Setomaa journalist and cultural activist, explains that "If some foreign politician needs to be impressed, we are invited as we are pretty and interesting." Despite these noble aims, however, the result has been an idealised image of Setomaa, whereby "from the outside we look better than we actually are" (Ragnar, entrepreneurship and culture). In a context where regional policy is increasingly neoliberalised, this can result in a situation "where politicians think that you are doing so well anyway that they don't need to support you any longer" (Tõnis, politics and culture), or the tremendous efforts in the region are co-opted as a best-practice example by a neoliberal discourse that celebrates local responsibility and a retreat of the state (Plüschke-Altof & Grootens, 2019).

On the one hand, the rural authenticity elite in Setomaa is also critical of such co-optation attempts, pointing out how local activists have "terribly overburdened themselves" (Greeta, journalism and culture) and lack time and financial resources to tackle material problems (Jaagup, entrepreneurship and community initiative), provocatively inviting others to "try to eat a song" (Erki, resident with Seto origin) and calling for a more redistributive regional policy (Märt, politics and culture). On the other hand, however, they are perpetuating the neoliberal discourse of self-responsibility by blaming other locals for persistent difficulties in the region despite all the efforts. This often takes the form of a division into the active and the non-active and hierarchies related to this (see also Annist, 2011: 280–283). Cultural activists Greeta and Toomas, for example, criticise the non-active as "very demanding" while at the same time not willing "to take on any kind of responsibility". In line with the research on territorial stigmatisation, this simultaneous appreciation of activeness and a lack of understanding for the non-active is often accompanied by a depiction of social pathologies of the latter, who are portrayed as development-resistant alcoholics who "stare at the bottom of the bottle", welfare abusers wanting "only social welfare payments" (Diana and Stiina, youth workers) or people "stuck in the Soviet era and with close to zero personal initiative levels" (Toomas, culture and politics). Indeed, there is an association between the population so described and the assumption of them living in one of the Soviet-era apartment blocks in the villages.

Thus, there is an interesting ambiguity that mirrors how local actors are also deeply embedded in neoliberal discourses. While "active Setos" appear to take a critical stance towards local responsibilisation by emphasising the price they pay for taking on responsibility for regional development, they also reproduce these discourses by setting themselves as positive role models of leadership and activeness – and, even more crucially here, by blaming and responsibilising exactly those among the local population who do not partake in the new development and cannot live up to these roles.

4.5.2 Dispossession as a result of a neoliberalisation of rural authenticity

However, the locals are not merely passive bystanders, subjected to such blame. They also actively criticise the process of "setotamine" (translatable very approximately as "Setoification"): that is, excessive delivery of Seto culture. "All the decisions are very Seto-centred," says Anne, a summer dweller in the region, "[s]o that…whatever is happening, it has to be in the name of having more Seto stuff, that Setos would be promoted more." Kaie (resident and self-declared Seto) contemplates the spatial exclusion resulting from the rural authenticity regime: "In my village [not seen as a cultural centre] there is nothing. We have been left like an orphan, behind. They built a new bus stop here, to A. village [the cultural hub], it's because of tourists. Ours can stay as it is because not many tourists come there." And Peeter (resident), who identifies as Estonian despite mentioning some potentially Seto grandparents, is particularly indignant about the possibilities for going out: "Yeah, some sort of…social events would be necessary here, not – I don't know… again some…you hear that Seto Leelo… [rolls his eyes]."

This critique mirrors the way in which those "taking care of Seto things" have defined a development path through rural authenticity that not only equates Setomaa with Seto people, but also defines Seto identity through cultural activism. The persistent link between Setomaa as a region and culturally active Seto people runs the risk of excluding those who do not feel connected to a culturally defined regional identity (Plüschke-Altof, 2018). Hence, basing the rural authenticity regime on the premise of Seto activism might result not only in the marginalisation of those local people who do not share Seto identity, but also of those Seto people who do not define themselves as active Setos and "would just like to live in peace" (Greeta, journalism and culture). At a gathering of a family identifying as Seto and Estonian, but not part of the active Seto elite, for example, it was debated whether "they feel part of this *thing*" (Jüri, son), referring to the Seto Kingdom Day, and how good they will feel there. Such discussions not only reveal different levels of commitment to Seto identity and its cultural expression, they also point at the increasing unease triggered by an established hegemony, complete with hierarchies of authenticity. These hierarchies do not remain "neutral", or even just a judgement in the daily interaction between people, but have very real consequences, as seen in the losses described by Peeter and Kaie.

A further dimension of such dispossessive processes unfolding in relation to authenticity-making has to do with funding opportunities. The majority of those applying to the above-mentioned funds "tend to be the same active Setos" (Mart, resident, self-declared Seto). Their success is multi-layered: partly related to the national preference for Seto cultural activities, and partly to the skills of those well established in the – inevitably Seto – cultural life. Finally, the links they have in the area to other active individuals and groups, united in their common goal of promoting and enhancing Seto culture, are decisive in ensuring financial support for such groups (Annist, 2005).

This is in stark contrast to the lack of such capacities or central uniting interests of the rest of the inhabitants in the region. To understand how this has evolved, Bourdieu's (Bourdieu, 1986) theory of forms of capital combined with the concept of dispossession is instrumental. The locals well embedded in Seto culture have gained symbolic capital with the increasing recognition and admiration of Seto culture. This has had an effect on their economic capital, but, through the process of conversion, also their social capital. And vice versa: loss of any such form of capital – in other words, dispossession – triggers a loss of other forms. The symbolic dispossession starts from a (inter)national recognition of the Setos, and evolves into a sort of "accumulation by dispossession", as the admired cases are juxtaposed with the rest, now seen in negative light as without an identity, even a Soviet one. These locals, now increasingly without cultural outlet, are then deemed inactive, insufficiently embedded in the particular heritage Seto authenticity. They are described as culture-less, wasted cases. One mother reported that, when she decided not to send her child to the local Seto school but chose another across the border in Võru county, she was criticised on social media by some local politicians, who suggested that "it's actually best if *mums like that* send their kids to other schools" (Kaie, resident, self-declared Seto [emphasis by the interviewee]).

Whilst Seto authenticity has triggered a new lease of social, economic and symbolic opportunities, as well as a common purpose for those involved, for others the post-Soviet reality of eroded and fragmented existence is not eased, but rather, aggravated. Lack of work opportunities in the new "Seto industry" means these locals experience regular, often longer-term mobility for work, as is characteristic of many in the rural periphery. Their access to having a say in local matters, in having a sense of belonging, has lessened. These changes also further erode their links to others, even if in similar circumstances. Lacking a common purpose, the rest of the locals have for years existed in a socially fragmented and symbolically rejected state – their defining feature within the local vicinity. This is further converted into economic dispossession: for those individuals unable to find local opportunities, but also for causes not centred on Seto authenticity.

The qualms expressed by such locals barely make it to public or political consciousness. Indeed, attempts to put forward a realistic picture of Seto life, warts and all, may further undermine the groups experiencing a different, troubled and challenging reality. In fact, expressing such qualms is locally framed not as reflecting important local issues, but rather as a reflection of the failures of such people themselves and/or their political representatives. Qualms are in fact more generally marginalised: as rural, and, in particular, peripheral regions' symbolic position in the country is low overall (Plüschke-Altof, 2016), airing negative issues is seen as an attempt to undermine the efforts of the local and/or cultural elites to build a positive image of local life. Those who wish to address persistent problems may even be treated as "traitors" or "trouble-makers" attempting to destroy the positive image others have worked so hard for (see also Plüschke-Altof & Grootens, 2019). Hence, the successes the Seto have benefited from have not only failed to extend to other locals, but expressing qualms about the selectiveness of the advantages further contributes to their negative image.

Those processes are shaped and aggravated by the emerging class relations in both the country and the region. The expression of Setoness is a marker – a distinction, to borrow Bourdieu (1984) – which only increases the demand to express authenticity, with gradually increasing strictness and attention to detail. Whilst this indeed leads to clarity of identity, it comes at a cost. The expectations in terms of living quarters, costumes worn, participation and preference for activities work as a distinguisher, separating those who can afford and are interested in honing this authenticity from those who are unable or uninterested in doing so. Such boundary-making when carving out who is and who is not Seto plays an increasing part in some locals' dislike of the Seto prominence. It also furthers the fragmentedness of the local social life.

To sum, from this focus on a particular rural authenticity, a new class structure starts to emerge, along with a new "cultural heritage elite" that promotes a particular ideal of Seto authenticity. Some of the rest of the population – both locals with Seto and with Estonian backgrounds – have been left behind by such developments. On the one hand, their particular display of "locality" is deemed to be inadequate or hybridised – that is, associated with both physical as well as supposedly mental remnants of the Soviet era. On the other hand, their ability to display their "sociality" is considered to be equally wanting as they abstain from Seto activities and fail to apply for funds made available in the local region, although often earmarked for Seto activities. As funding has become increasingly important for most group activities locally, inability to secure such funding is seen to further demonstrate the inadequacy of some locals. New classes are emerging locally as a result of these processes, with those at the bottom struggling to gain economic, social and symbolic capital as they are associated with negative (peripheral) rather than positive (authentic) rurality. These new power relations in the countryside lead to an accumulation of dispossessions.

4.5.3 Populism of those dispossessed by the rural authenticity regime

Signs that dispossessions might have something to do with the rise and success of populist parties have been emerging in East European countries for several years now (see Kalb, 2009a, 2009b; Pilkington, 2016; Narotzki, 2019). The Seto region, along with the surrounding Võru county (which has another, albeit not as powerful and mostly dialect-based minority alignment) can be identified as a populist hotspot: the support for the far-right party EKRE (Estonian Conservative People's Party) during the 2019 national elections was nearly 30%: the third highest in Estonia. Whilst there are other reasons for this support, the loss of local life chances for the population not aligned with the Seto cause is likely to have played an important part: "It's what the local social democratic government of the new [unified Seto] municipality chooses to do. They plough money into Seto model villages. So who would you vote for instead?" (Peeter, resident, self-declared Estonian with some Seto roots). Those left out of nation-making processes, and often tepid towards any political processes, are increasingly irritated by "the small segment getting money, sharing it between themselves and then getting all puffed up what great Setos they are" (Peeter). With the emergence of the populist party, he expresses some hope this "merely commercial, and tragicomically two-faced" cultural elite would be put in their place. The "Seto versus the rest" rift is also a remarkable example of the role of a peculiar kind of "white identity politics" wherein the nationally dominant Estonian identity has, with blessing from the national government and the Seto elite and aligned locals, succumbed to local rural authenticity making.

Grievances over the centrality of authenticity and heritage culture amongst the population not aligned with Setoness and thus being left behind peaked in the 2017/2018 period, during the Estonian municipality amalgamation reform. The political struggles against the development of a new, united Setomaa municipality in Estonia was the first time the locals not behind the authentic Seto cause were able to unite around anything. In many ways, their cause was practical and had to do with the local centres – for instance, there tend to be better bus connections within a municipality but not across its borders. With the change in administrative borders, those in the new local peripheries were concerned about their ability to access their usual nearby but now cross-municipality centres. Their failure, and the resulting establishment of the Setomaa municipality, may have been the final push towards such large numbers choosing the populist party representing Estonian-ness as the sacred goal of the country.

4.6 Conclusion

The chapter discusses the multiple dispossessions arising from local, national and European Union level support structures and funding to promote authenticity and management of heritage culture for rural development in the context of a post-socialist neoliberalisation of regional politics. It shows how the reliance on Seto authenticity establishes a discursive hegemony linking authentic Setoness to active performance in the field of cultural heritage and thereby excludes not necessarily externals and newcomers, but those locally marginalised people unable or unwilling to properly perform "active Setoness". Moreover, by idealising Setomaa and (a clearly defined group of) active Seto inhabitants in order to foster regional development in times of regional competitiveness, it disguises (and even actively discourages) discussing persistent (structural) problems in the region and therefore faces the danger of being co-opted by a neoliberal discourse propagating local responsibility for regional development and justifying the roll-back of state social services. These processes hit particularly hard those locals who have experienced various forms of dispossession due to the post-socialist changes, the fragmentation of the peripheral communities and the increased need for mobility. As neoliberal expansions have created surplus populations (Salemink & Rasmussen, 2016), the benefits of both market and social liberalism have passed them by. Whilst these processes are not specific to any locales, it can be expected that the local qualms and juxtapositions with emerging local identity-based advantages would make siding with national populism more likely.

Setomaa consequently resembles a case wherein the appeal of populism to the local residents is not so much the result of a targeted use of rural authenticity and idyllic images by populist movements. Instead, it might be the result of dispossessions created by a neoliberal employment of rural authenticity in times where the responsibility for regional development is gradually shifting towards the regions themselves – irrespective of their resources and capacities to do so. The populism of the dispossessed can thus be interpreted as the expression of dissatisfaction with the rural authenticity development path and the exclusions and marginalisation it entails, which are often underacknowledged in a context where a reliance on rural authenticity is seen as without alternatives due to the increasing retrenchment of a neoliberalising state. The chapter therefore seeks to open the debate on the role of neoliberal (regional) policies as the often overlooked link between rural authenticity and populism.

Acknowledgement

The research presented here received funding from the People Program (Marie Curie Actions) of the European Union's Seventh Framework Program FP7/2007-2013/ under REA grant agreement n°607022 "Socio-economic and Political Responses to Regional Polarization in Central and Eastern Europe"; MSVOI16094R "PROMoting youth Involvement and Social Engagement: Opportunities and challenges for 'conflicted' young people across Europe"; MJD450 "Mapping the migratory careers of transnational Estonians"; IUT34-32 "Cultural heritage as a socio-cultural resource and contested field"; and PRG908 "Estonian Environmentalism in the 20th century: ideology, discourses, practices".

Notes

1 In this chapter we will use "Seto region", "Seto county" and the Estonian version "Setomaa" interchangeably.
2 While there are up to 22,000 self-declared Seto speakers, only a minority of them lives in Setomaa. Among the 3,280 Seto county inhabitants are Seto speakers as well as non-speakers.

References

Annist, A. (2005) The worshippers of rules: Defining the right and wrong in local project applications in Estonia. In Mosse, D. and Lewis, D. (eds), *The aid effect: Giving and governing in international development*. Pluto Press: London, pp. 150–170.

Annist, A. (2011) *Otsides kogukonda sotsialismijärgses keskuskülas*. Arenguantropoloogiline uurimus. ACTA Universitatis Tallinnensis: Tallinn.

Annist, A. (2013) Heterotopia and hegemony. Power and culture in Seto country. *Journal of Baltic Studies* 44(2): 249–269.

Annist, A. (2015) Formal crutches for broken sociality. In Morris, J. and Polese, A. (eds), *Informal economies in post-socialist spaces: Practices, institutions and networks*. Palgrave Macmillan: London, pp. 95–113.

Annist, A. (2017) Emigration and the changing meaning of Estonian rural life. In Tammaru T (ed.), *Estonian human development report: Open Estonia*. Eesti Koostöö Kogu: Tallinn, pp. 247–254.

Annist, A. (2018) Struggling against hegemony: Rural youth in Seto country, Estonia. Case study report within Manchester University PROMISE Project, Horizon2020. www.promise.manchester.ac.uk/wp-content/uploads/2019/03/Individual-case-study-Estonia-rural-youth-in-Seto.pdf

Annist, A. (2022) Post-socialism as an experience of distancing and dispossession in rural and transnational Estonia. *Critique of Anthropology. Special Issue: The anthropology of postsocialism: Theoretical legacies and conceptual futures* (forthcoming).

Ashworth, G. J. (2009) The instruments of place branding. How is it done? *European Spatial Research and Policy* 16(1): 9–22.

Baumann, C. (2018) *Idyllische ländlichkeit*. Eine Kulturgeographie der Landlust. Bielefeld: Transcript Verlag.

Bendix, R. (2009) Heritage between economy and politics: An assessment from the perspective of cultural anthropology. In Smith, L. and Akagawa, N. (eds), *Intangible heritage*. Routledge: London, pp. 253–269.

Bohle, D., and Greskovits, B. (2012) *Capitalist diversity on Europe's periphery. Cornell studies in political economy*. Cornell University Press: Ithaca, London.

Bourdieu, P. (1984) *Distinction: A social critique of the judgement of taste*. Routledge and Kegan Paul: London

Bourdieu, P. (1986) The forms of capital. In Richardson, J. (ed), *Handbook of theory and research for the sociology of education*. Greenwood: New York, pp. 241–258.

Bristow, G. (2005) Everyone's a "winner": Problematising the discourse of regional competitiveness. *Journal of Economic Geography* 5(3): 285–304.

Bunten, A. C. (2004) Commodities of authenticity: When natives consume their own "Tourist Art". In Welsch, R., Venbrux, E. and Scheffield, R. P. (eds), *Exploring world art*. Waveland Press: Long Grove, pp. 317–336.

Bunten, A. C. (2008) Sharing culture or selling out? Developing the commodified persona in the heritage industry. *American Ethnologist* 35: 380–395.

Creed, G. W. (2011) *Masquerade and postsocialism: Ritual and cultural dispossession in Bulgaria*. Indiana University Press: Bloomington.

Escallon, M. F. (2017) The formation of heritage elites: Talking rights and practicing privileges in an Afro-Colombian community. In Silverman, H., Waterton, E. and Watson, S. (eds) *Heritage in action*. Springer: Cham, pp. 63–74.

Halfacree, K. (2006) From dropping out to leading on? British counter-cultural back-to-the-land in a changing rurality. *Progress in Human Geography* 30(3): 309–336.

Halfacree, K. (2007) Trial by space for a "radical rural". Introducing alternative localities, representations and lives. *Journal of Rural Studies* 23(2): 125–141.

Hann, C. (2011) Moral dispossession. *InterDisciplines* 2: 11–37.

Harrison, R. (2013) *Heritage: Critical approaches*. Routledge: Oxford.

Harvey, D. (2004) The "new" imperialism: Accumulation by dispossession. *Socialist Register* 40: 63–87.

Harvey, D. (2007) *A brief history of neoliberalism*. Oxford University Press: Oxford.

Hayes, M. (2020) The coloniality of UNESCO's heritage urban landscapes: Heritage process and transnational gentrification in Cuenca, Ecuador. *Urban Studies* 57(15): 3060–3077.

Kalb, D. (2009a) Conversations with a Polish populist: Tracing hidden histories of globalization, class, and dispossession in postsocialism (and beyond). *American Ethnologist* 36(2): 207–223.

Kalb, D. (2009b) Headlines of nationalism, subtexts of class: Poland and popular paranoia, 1989–2009. *Anthropologica* 51(2): 289–300.

Kašková, M., and Chromý, P. (2014) Regional product labelling as part of the region formation process. The case of Czechia. *AUC Geographica* 49(2): 87–98.

Kasmir, S., and Carbonella, A. (2008) Dispossession and the anthropology of labor. *Critique of Anthropology* 28(5): 5–25.

Kavaratzis, M., and Ashworth, G. (2015). Hijacking culture. The disconnection between place culture and place brands. *Town Planning Review* 86(2): 155–176.

Kay, R., Shubin, S., and Thelen, T. (2012) Rural realities in the post-socialist space. *Journal of Rural Studies* 28(2): 55–62.

Külvik, H. (2014) *Setomaa: Unique and genuine*. Värska: Seto Instituut.

Kuutma, K. (2019) Afterword: The politics of scale for intangible cultural heritage – identification, ownership and representation. In Lähdesmäki, T., Zhu, Y. and Thomas, S. (eds), *The politics of scale: New directions in critical heritage studies*. Berghahn Books: Oxford, pp. 156–170.

Kuutma, K., and Annist, A. (2020) Home and heritage out of place: The disjunction of exile. *International Journal of Heritage Studies* 26(10): 942–954.

Lang, T., Burneika, D., Noorkoiv, R., Plüschke-Altof, B., Pociūtė-Sereikienė, G., and Sechi G. (2021) *Socio-spatial polarisation and policy response – future perspectives for regional development in the Baltic States*. European Urban and Regional Studies (online first DOI: 10.1177/09697764211023553).

Lang, T., and Görmar, F. (eds) (2019) *Regional and Local Development in Times of Polarisation. Re-thinking Spatial Policies in Europe*. Palgrave Macmillan: Houndmills.

Lauristin, M., and Vihalemm, P. (2009) The political agenda during different periods of Estonian transformation: External and internal factors. *Journal of Baltic Studies* 40(1): 1–28.

Loewen, B. J. (2018) *Towards territorial cohesion? Path dependence and path innovation of regional policy in central and eastern Europe*. University of Tartu Press: Tartu.

Lõuna, K. (2003) *Petserimaa: Petserimaa integreerimine Eesti Vabariiki 1920–1940*. Tallinn: Eesti Entsüklopeediakirjastus.

Mazzarella, W. (2019) The anthropology of populism: Beyond the liberal settlement. *Annual Review of Anthropology* 48: 45–60.

Narotzki, S. (2019) Populism's claims: The struggle between privilege and equality. In Kapferer, B. and Theodossopoulos, D. (eds), *Democracy's paradox: Populism and its contemporary crisis*. Berghahn Books: New York and Oxford, pp. 97–112.

Nugin, R. (2014) "I think that they should go. Let them see something": The context of rural youth's out-migration in post-socialist Estonia. *Journal of Rural Studies* 34: 51–64.

Palang, H., Semm, K., and Verstraete, L. (2009) Time borders: Change of practice and experience through time layers. *Journal of Borderland Studies* 24(2): 92–105.

Pilkington, H. (2016) *Loud and proud: Passion and politics in the English Defence League*. Manchester University Press: Manchester.

Plüschke-Altof, B. (2016) Rural as periphery per se? Unravelling the discursive node. *Sociální studia / Social Studies* 13(2): 11–28.

Plüschke-Altof, B. (2017) The question of responsibility: (De-)peripheralising rural spaces in post-socialist Estonia. *ESR&P* 24: 59–75.

Plüschke-Altof, B. (2018) Re-inventing Setomaa: The challenges of fighting stigmatisation in peripheral rural areas in Estonia. *Geographische Zeitschrift* 106(2): 121–145.

Plüschke-Altof, B., and Grootens, M. (2019) Leading through image making? On the limits of emphasising agency in structurally disadvantaged rural places. In Lang, T. and Görmar, F. (eds) *Regional and local development in times of polarisation*. Palgrave Macmillan (New Geographies of Europe): Singapore, pp. 319–341.

Plüschke-Altof, B., Loewen, B., and Leetmaa, K. (2020) Increasing regional polarisation in Estonia. In Sooväli-Sepping, H. (ed), *Estonian Human Development Report: Spatial Choices for an Urbanised Society*. Eesti Koostöö Kogu: Tallinn, pp. 44–55.

PoSCoPP Research Group (2015) Understanding new geographies of Central and Eastern Europe. In Lang, T., Henn, S., Sbignev, W., and Ehrlich, K. (eds), *Understanding geographies of polarization and peripheralization*. Palgrave Macmillan: Houndmills, pp. 1–24.

Pungas, L. (2017) Soziale Kosten der ökonomischen Transformation in Estland: Der Preis des Wachstumsparadigmas. *Ost Journal* 1: 10–19.

Raagmaa, G., Masso, J., Reidolf, M., and Servinski, M. (2012) Empowering people and enterprises with strong cultural and territorial identity. A case study of Setomaa, Estonia. In Kinnear, S., Charters, K. and Vitartas, P. (eds), *Regional advantage and innovation*. Springer: Wiesbaden, pp. 233–254.

Sahlins, M. (1999) Two or three things that I know about culture. *The Journal of the Royal Anthropological Institute* 5(3): 399–421.

Salemink, O., and Rasmussen, M. B. (2016) After dispossession: Ethnographic approaches to neoliberalization. *Focaal* 74: 3–12.

Smith, L. (2006) *Uses of heritage*. London: Routledge.

Storie, J., Chenault, E., Külvik, M., and Bell, S. 2020. When peace and quiet Is not enough: Place-shaping and the role of leaders in sustainability and quality of life in rural Estonia and Latvia. *Land* 9(8)/259. https://doi.org/10.3390/land9080259

Valk, A., and Särg, T. (2015) Setos' way to manage identities and well-being. *Nationalities Papers* 43(2): 337–355.

Vöörmann, R. (2005) Tööturu ohud: Tööhõive vähenemine ja töötuse suurenemine. In Raitviir, T. and Raska, E. (eds), *Eesti edu hind. Eesti sotsiaalne julgeolek ja rahva turvalisus.* Eesti Entsüklopeediakirjastus: Tallinn, pp. 74–77.

Winter, T. 2015 Heritage and nationalism: An unbreachable couple? In Waterton, E. and Watson, S. (eds), *The Palgrave handbook of contemporary heritage research.* Palgrave Macmillan: London, pp. 331–345.

Woods, M. (2013) Regions engaging globalization: A typology of regional responses in rural Europe. *Journal of Rural and Community Development* 8(3): 113–126.

Wrobel, R. M. (2013) Estland. In Heydemann, G. (ed), *Vom Ostblock zur EU: Systemtransformationen 1990–2012 im Vergleich,* 1. Aufl. Vandenhoeck & Ruprecht: Göttingen, pp. 17–46.

5   The Minister's tears and the strike of the invisible

The political debate on the "regularisation" of undocumented migrant farm labourers during the Covid-19 health crisis in Italy

Domenico Perrotta

5.1 Introduction

In mid-May 2020, while Italy was slowly emerging from the Covid-19 health crisis, the government launched the "Decreto Rilancio" (the "Restart Decree"): more than 300 pages containing a great number of measures aimed at fostering the restart of the Italian economy.[1] On May 13, the prime minister and other members of the government presented the decree in an official press conference. The then Minister of Agriculture, Teresa Bellanova, briefly introduced one of the most discussed measures of the decree: the "formalisation of job relationships" ("emersione dei rapporti di lavoro") – namely, the "regularisation" programme for foreign citizens living in Italy without a permit of stay. Most of them, it is said, were irregularly employed as agricultural labourers, in conditions of severe exploitation, and were dramatically exposed to the epidemic. In describing this measure, the Minister could not hold back her tears:

> Since today, thanks to this government's decision, the invisible will be less invisible. Those who have been brutally exploited in the fields [...] won't be invisible, because they will be able to obtain a permit of stay for labour reasons, and we will help them to be persons that regain their identity, their dignity.
>
> (Document 5, see Appendix)

Her emotion was due, she explained, to the relevance of this programme to her life history and origins: as a young girl, as she often underlines, she was employed as a farmworker in Puglia, her region of origin; to confront labour exploitation she engaged as union activist before becoming a politician. For this reason, she affirmed, she feels the dramatic conditions of contemporary farm workers as very close to her own experience.

From the end of February to May 2020, Italy was hit by a health crisis and related lockdown. In these months, the political debate on agriculture, which is usually focused on the magnification and defence of "Made in Italy" food,

DOI: 10.4324/9781003091714-6

appeared to strongly acknowledge the structural role of foreign workers in the Italian fields. A segment of such workers decided to go back to – or were obliged to remain in – their home countries, especially in Eastern Europe; other workers – those who did not have either a permit of stay or a labour contract – were not able to move from their dwelling places to the fields, or from one Italian region to another to follow the seasonal labour demand. The two major Italian farmers' organisations, Coldiretti and Confagricoltura, announced the risk of a relevant workforce shortage, estimated respectively at 370,000[2] and 200,000 workers,[3] and requested urgent measures, such as "green corridors", to facilitate the transnational mobility of Eastern European seasonal labourers.

While some farmers tried to solve the problem by themselves,[4] and the government extended expiring permits of stay, the public debate increasingly focused on the opportunity of an "amnesty" for undocumented migrants. In the preceding months, the Italian government had already considered such a measure because of the great number of undocumented migrants in the national territory;[5] however, the health crisis now made the measure urgent, with the aim of securing the presence of these workers in the fields.

In this chapter, I do not intend to discuss the regularisation programme[6] (for a first assessment, see Palumbo & Corrado, 2020) and its outcomes.[7] Rather, I focus on the political debate surrounding this decree, in particular between 16 April 2020, when the Minister of Agriculture officially announced it to the Italian parliament, and 21 May 2020, when – after the Decree was approved by the government – a small grassroots labour union, the USB (Unione Sindacale di Base), organised a strike of migrant farm labourers to protest against the conditions for accessing the regularisation.

This political debate appears relevant to an understanding of the changing policies and politics of agriculture and food in Italy. At least in part, it seems to redefine some widespread representations of the Italian agri-food system, with tension between the claims of the "excellence" of "Made in Italy" food – recently used in a nationalist way by radical-right populist parties (Iocco, Lo Cascio, & Perrotta, 2020) – and the denouncements of the dramatic working and living conditions of migrant farmworkers. After a short discussion of the representations of the Italian agri-food system (Section 5.2), the analysis will focus on the three main actors of this political debate: the Minister of Agriculture, Teresa Bellanova, a member of the centre-left party "Italia viva" (lead by former Italian prime minister Matteo Renzi); the head of the opposition and leader of the populist radical-right Lega party, Matteo Salvini; and Aboubakar Soumahoro, leader of the USB and a grassroots union organiser. In Section 5.3, I propose a discourse analysis inspired by Laclau and Mouffe's (1985) discourse theory.[8]

If, with Mudde and Rovira-Kaltwasser (2017, p. 1), we define populism as a thin-centered ideology that considers society to be ultimately separated into two homogeneous and antagonistic camps, 'the pure people' versus 'the corrupted elite', and which argues that politics should be an expression of the *volonté générale* (general will) of the people, the analysis of this debate shows that all three discourses involve elements of populism, even if they are differently articulated within other "moments"[9] of the respective political positions. Salvini clearly expresses a form of radical-right populism (articulating populism with nativism and authoritarianism), while Soumahoro seems to propose a form of left-wing populism that aims to unite different social classes (farmworkers, peasants, urban consumers, etc.) against an "economic elite" (in this case, the big retail chains). Bellanova – whose centre-left political position can be considered as neoliberal – proudly proclaims her popular origins as a farmworker. In this sense, not only do elements of populism cross all the positions in the political debate on agriculture and farm labour in Italy, but, as the three actors seem to argue, only by representing him/herself as a member of the people one can legitimately discuss Italian agriculture.

For this chapter, I considered a number of documents produced by these three political actors on the issue of the regularisation of undocumented migrant farm labourers during the months of April and May 2020. In particular, I quote three kinds of texts: speeches and press conferences of Teresa Bellanova and Matteo Salvini at the Italian Parliament (documents 1, 2, 5, 9: see Appendix); interviews in newspapers and TV news reports, and participations in TV talk shows (documents 3, 4, 6, 8); and press releases (document 7).

5.2 Hegemonic and counter-hegemonic representations of Italian agri-food systems and farm labour

Italy is known as a "gastronationalist" country (DeSoucey, 2010; Benasso & Stagi, 2019) in the domains of both policy and politics. On the one hand, Italy is the European country with the highest number of food products and wines with protected denominations of origin (306 and 525, respectively, as of September 2020); on the other hand, since the 1990s, in the Italian debate on food and agriculture a "Made in Italy food consensus" has been reached by the most important public and private actors of the Italian agri-food system on the fact that *all* Italian food should be considered "quality" food (Brunori, Malandrin, & Rossi, 2013): this "quality" is associated to artisanal more than industrial production, to the Italian "national identity" as well as to local productions and, in some versions, to biodiversity and family farming. The origins of this "quality turn" can be dated back to the 1970s, with the birth of the organic agriculture movement, and the 1980s, when Carlo Petrini founded Slow Food. In the 1990s, after some food-safety scandals, the consensus around these issues grew, not only among grassroots groups of consumers, with the movement of the "gruppi di acquisto solidale" (solidarity purchase groups; Grasseni, 2013), but – most importantly – among political parties, the major farmers' organisations, agro-industrial corporations and retail chains. The Expo 2015 in Milan probably represented the apex of this view of Italian agriculture (Howard & Forin, 2019).

The "quality turn", the "Made in Italy food consensus" and the hegemonic representation of Italian agriculture that depicts Italian rural areas as the regions of origin of thousands of "authentic" and "traditional" food products have not prevented the diffusion of processes such as the rapid decrease in the number of Italian farms, the growth of the average farm dimensions, the industrialisation of Italian agriculture or the retailisation of food distribution. Fonte and Cucco (2015) argue that the Made in Italy food consensus can be seen as a process of conventionalisation of those movements – such as the movement for organic agriculture and Slow Food – which originally represented forms of radical opposition to the dominant agri-food system.

What is more, this representation of *Italian* agriculture hides the structural role of hundreds of thousands of non-Italian farm labourers, who often experience dramatic working and living conditions, especially in the Southern coastal plains (Corrado, de Castro, & Perrotta, 2016; Corrado, Lo Cascio, & Perrotta, 2018). Nonetheless, as argued by Howard and Forin (2019) in their study on the production and representation of the Italian industrial tomato, there is a "counter-hegemonic imagery" of Italian agriculture that they define as "sensationalized brutality", which over-emphasises "the migrant worker's suffering to the extent that the hegemonic image erases his presence" (p. 585). This counter-hegemonic representation is constructed mainly by civil society groups, such as the main labour unions, as well as by news reports. It underlines key words such as exploitation, slavery and *caporalato* (a system of informal and illegal farm labour intermediation which is widespread in Italian intensive agriculture: see Perrotta & Sacchetto, 2014; Avallone, 2016; Salvia, 2020) and holds farmers and "big businesses" as directly or indirectly responsible for labour exploitation, as well as blaming the Italian government for its "inaction". It is what Howard and Forin (2019) call a form of "modern abolitionism", which, in their opinion, is as inaccurate as the "hegemonic" narrative. Iocco et al. (2020, p. 9) note that "the 'fight' against the exploitation of agricultural labourers became a priority for national governments and local administrations and the idea that such exploitation is not acceptable is now largely shared by the Italian public opinion". In the 2010s, the centre-left Italian governments, and especially their Minister of Agriculture, Maurizio Martina, attempted to make a "synthesis" of the hegemonic and counter-hegemonic representations of Italian agriculture, through a "humanitarian" approach to the issue of migrant labourers' living and working conditions:

> It is precisely here that the discourse on Italian "fair" and "good" agriculture, and the discourse of the "humanitarian" approach to migrant (agricultural labour) crisis found an original but precarious articulation in the 2010s. On the one hand, centre-left governments responded to the crisis of small- and medium-scale farming deploying a re-energised rhetoric focused on the Made in Italy "quality" food, interestingly, expanding such discourse to the struggle against the exploitation of migrant

workers. On the other hand, the same governments responded to the "migrant crisis" with a "humanitarian" approach in the domain of reception, even towards agricultural labourers.

(Ibid., p. 11)

On the other side of the political spectrum, over the same years the centre-right coalition has been hegemonised by the "new" Lega[10] lead by Matteo Salvini, whose leadership substituted Silvio Berlusconi's "old" Forza Italia. The new Lega can be considered a radical-right populist party, with its characteristics of nativism, authoritarianism and populism (Mudde, 2007). Iocco et al. (2020) analysed Lega's representation of Italian agri-food, taking into consideration its political discourse when the party was in opposition as well as during its short government experience with the other Italian populist party, the Movimento 5 Stelle (Five Star Movement), from June 2018 to August 2019. They argued that Salvini's Lega re-articulated the "hegemonic" representation of Made in Italy "quality" food in a relatively new nativist and corporatist discourse, in which food and agriculture are represented as central to the "Italian" cultural identity. They described how the features of nativism, authoritarianism and populism are articulated in Lega's political discourse on agriculture and food; moreover, they analysed Lega's economic approach to the agri-food system, emphasising its neo-mercantilism – which is not new in EU agri-food policies (Potter & Tilzey, 2005) – and corporatism.

The political discourse of the centre-left governments (2013–2018) and the Lega's government experience (2018–2019) seem to differ partially in their respective articulations of the topic of (the exploitation of) migrant farm labourers: namely, the "humanitarian" approach of the centre-left and the "authoritarian" Lega approach. Salvini represented the issues of *caporalato* and labour exploitation as public order problems, associated with mafia clans and illegal immigration, to be severely punished, while he depicted the majority of Italian farmers as "honest". Concordant with this view, when serving as the Minister of Internal Affairs, Salvini sent bulldozers to evacuate some rural informal ghettos in which seasonal labourers of Sub-Saharan African origin often find a precarious roof during the harvest seasons in the Southern regions of Calabria and Puglia, and claimed that these interventions re-established the rule of law (Iocco et al., 2020).

This kind of action – such as the eviction of informal settlements – had been taken by the centre-left governments as well: both right- and left-wing governments share the idea that the *caporalato* must be treated as a criminal offence and punished with jail; this idea was embodied by Law 199-2016 (issued by a centre-left government), which punishes both the informal brokers and the farmers that hire labourers in a condition of exploitation. Nonetheless, while on the one hand in the centre-left political discourse such penal measures and the eviction of informal ghettos are seen as necessary to "save" migrant labourers, represented as the victims of their exploiters, on the other the right-wing political discourse represents such labourers as "illegal migrants", and thus part of the problem.

5.3 The health crisis and the debate on the "regularisation of the invisible" migrant workers

To date, the political measures intended to address and resolve the dramatic issue of the working and living conditions of migrant (especially seasonal) labourers in Italian intensive agriculture have not reached their goals. Notwithstanding the judicial inquiries against the *caporali* and employers, which followed Law 199/2006, and notwithstanding the evictions of some informal ghettos, the majority of migrant labourers who move from one region to another following the seasonal harvests have no alternative to the ghettos and the *caporali* if they want to find a (precarious) roof and (under-paid) employment.

Indeed, over the last few years the number of migrant labourers seeking employment in these areas has probably increased, partly because of the high number of denials of asylum applications (see endnote 5) and partly due to the difficulties in finding a job in other sectors and in other regions of Italy.

For these reasons, the health crisis connected to the Covid-19 pandemic raised a number of concerns. As noted in the Introduction, farmers' organisations denounced the risk of labour shortages in agriculture because of the restrictions on workers' mobility, both from abroad and among the Italian regions. Additionally, migrant workers, especially those living in the ghettos and informally employed in agriculture, found it increasingly difficult to move from their dwelling places to the fields, as well as from one region to another at the end of the harvest season, and they were afraid of losing their prospective (albeit meagre) earnings. Furthermore, the sanitary conditions in the ghettos – with no water or electricity, and the impossibility of respecting "social distancing" measures – created fears of an uncontrolled spread of the virus in these places.

It was precisely this situation that prompted Minister of Agriculture Teresa Bellanova, in a discourse to the Italian Parliament on April 16, to announce the government's intention to launch a regularisation programme for undocumented migrants. Of course, not all undocumented migrants in Italy work in agriculture and live in the rural ghettos. We can estimate a presence in Italy of 3–400,000 undocumented migrants (while the minister estimated a number of 600,000); this figure is roughly equivalent to the total number of migrant workers who signed at least one formal contract in agriculture (370,000 in 2017; see Crea, 2019).[11]

However, the political debate on undocumented migrants during the health crisis was dominated by the situation of farm labourers, for at least two reasons. First, because of their hyper-visibility: as noted by Howard and Forin (2019), the "modern abolitionist" counter-hegemonic imagery on Italian agriculture is widespread in the Italian mass media and news reports; the Sub-Saharan African farmworkers living in the ghettos and working in the fields in Southern Italy are often the object of journalistic inquiries and newspaper articles. Second, because the risk of a workforce shortage in the fields during the health crisis was represented as a risk to the food security of the country.

Undocumented migrant labourers, in this situation, have sometimes been represented as "heroes" who, in the context of the pandemic, have continued to work in the fields to feed the Italian population. Similarly, another category of undocumented workers, namely domestic workers, were admitted to the regularisation programme for similar reasons: these (mainly female) workers were also presented as "heroes" who, during the health crisis and notwithstanding the risks to their own health, continued to take care of elderly Italian persons. Thus, similarly to the agricultural labourers, they are seen as deserving a permit of stay. The undocumented migrants who are (informally) employed in other economic sectors do not deserve such a reward.

In this context, and returning to the discourse analysis of the representations of agriculture and farm labour, the position of Teresa Bellanova in this debate can be analysed under two different axes: migration policies and agrifood policies. From the point of view of migration policies, she mixes two traditional approaches of the Italian centre-left coalition: a "utilitarian approach" to migration processes, and a "humanitarian" concern for the spread of the virus among migrant workers in the ghettos. The "utilitarian approach" considers foreign citizens as deserving of being admitted in the national territory only if and when they are helpful for the Italian economy. The primary reason that will convince the Italian parliament to adopt the "amnesty" measure resides in the (supposed) necessity of these (now undocumented) workers for Italian agriculture. I quote here some passages of her discourse to the Italian Senate:

> The food chain did not stop a single day of work. And yet critical issues and problems are on the agenda. [...] Excellent sectors, our ambassadors in the world markets, are in great trouble today. [...] In this context we must consider the dramatic warning launched by the companies and the farmers' organisations on the shortage of seasonal workers and the difficulty in the recruitment of the workforce, which can hopelessly harm the processing and harvest of the products. This can't happen. This doesn't have to happen. Not while in this country the number is growing of the citizens that are forced not to work and today hardly get fed. Not when it means to lose the work, the investments of the past years and the survival of the companies, the security of the employment of millions workers. This sector is so important, but it could not resist the impact of so many events. We can't allow it. [...] Farmers' associations describe a seasonal workforce shortage between 270,000 and 350,000 workers. We know that thousands of foreign workers, mainly from the East, that up to now were seasonally employed in our fields, went back to their home countries.
>
> (Document 1)

As she clearly argued in an interview to the Italian newspaper *Corriere della Sera*: "We are not doing a favour to immigrant citizens by giving them a permit of stay. We are simply recording the need of an additional workforce, which is not against unemployed Italian citizens" (Document 3).

Secondarily, from a "humanitarian" point of view, Bellanova argues, these workers are vulnerable and thus more exposed to the virus (and thus, she seems to imply, they can put the entire Italian population at risk):

> At the same time, other workers, invisible to the majority of the population, the so-called "irregulars", 600,000 according to the estimates, live in informal settlements, underpaid and often exploited in an inhuman way. Persons that in the majority of cases are already working on our territory, at the mercy – together with the companies where they work – of that criminality that we call *caporalato* and that means "mafia" for me. In the current situation, the conditions of such irregular workers are even more complicated and fragile, and these people even more exposed to the health risks and hunger. [...] The government has already adopted measures of health protection of migrant citizens who live in the irregular settlements, with the goal of preventing the spread of the Covid-19 infection in such a risky contexts. But it is not sufficient.
>
> (Document 1)

This "humanitarian" stance articulates the concerns for the epidemic with the traditional denouncements of the "slavery" in the Italian fields; this brings us to the analysis of Bellanova's representation of agriculture. In this domain, she tries to reconcile the (apparently irreconcilable) discourses we analysed in Section 5.2: the (hegemonic) emphasis on the "excellence" and "quality" of "Made in Italy" food and the (counter-hegemonic) denouncing of the severe exploitation of migrant labourers in the fields (Howard & Forin, 2019). On the one hand, she acknowledges the unquestioned quality of Italian agriculture: "We must implement the assets so that our farms and the food chain can do what they know most and that the world envies of us: quality food, the highest quality food" (Document 1).

On the other hand, she confirms her vision of migrant farmworkers as the victims of the *caporalato* system and the mafia clans, adding that: "Without ideologies and hypocrisies, dear colleagues, we have the duty of taking responsibility: either the State takes charge of the life of these people, or criminality will exploit it" (Document 1).

When, on May 13, the government presents the "Decreto rilancio", she claims authorship of the regularisation programme and explains her commitment to this controversial issue through the argument of her popular origins. Indeed, since September 2019, when she was appointed Minister of Agriculture, she had regularly underlined the fact that when she was younger she had worked as a farmworker and was severely exploited. That experience brought her to join the CGIL (Confederazione generale italiana del lavoro), the main Italian labour union, and to work as a union activist for many years before becoming a member of the Italian parliament.

In a press interview on May 14, she recalls this "exemplary" life history:

> My history is the history of a person who has known the brutal work in her childhood, has had a denied childhood, because children should have the right to play, and when, on the contrary, when you are a child you are forced to work, it is something that marks you for life [...] When you start fighting, already in your childhood, against these brutal forms of exploitation, you can't forget it when you arrive in such a place [the Ministry of Agriculture]; rather, if you are lucky enough to arrive in such a place, you must remember where you came from and who you have to give voice to, if he hasn't voice yet.
>
> (Document 6)

This history explains the Minister's emotion and tears while announcing the "amnesty" measure. At the same time, this argument has a populist flavour: the Minister proclaims her popular origins and, she says, even in the Ministry's rooms she cannot forget "where she comes from":

> The last point, Mr. President, that I want to recall, and that many can consider a secondary point, for me, for my history, is a fundamental point [she cries]. And I mean the [...] formalisation of job relationships. Since today, thanks to this government's decision, the invisible will be less invisible. Those who have been brutally exploited in the fields [...] won't be invisible, because they will be able to obtain a permit of stay for labour reasons, and we will help them to be persons that regain their identity, their dignity [...] From today, we can say that the State wins, because the State is stronger than the criminality, the State is stronger than the *caporalato*.
>
> (Document 5)

Moving to the analysis of Salvini's discourse in the debate on the "amnesty", we must note that it is concordant with his views on agriculture and on migration (Iocco et al., 2020), and with his usual "nativist" slogan "prima gli italiani" (Italians first). His first reaction to the announcement of the amnesty – during a press conference for his party in parliament – is the following: "The governments' proposal is to do an amnesty for 600,000 *clandestini*. Lega's proposal is to accompany students, *cassintegrati* [furloughed workers], unemployed, retired, Italians if possible, to get a job" (Document 2). Over the following weeks, he repeated this sentiment in tens of interviews, meetings and TV talk shows.

In such interventions, the leader of the centre-right coalition uses other recurrent elements of his view of agriculture: the defence of "Made in Italy" food production and of Italian farmers ("yesterday at the supermarket in Milan close to my house I found strawberries from Spain, citrus fruit from Spain, tomatoes from Belgium. No. We should reward the big retailers only if they distribute Italian products", Document 4); the necessity to cut "taxes and bureaucracy" for agricultural entrepreneurs; the idea that the large majority of "honest" and "worthy" Italian farmers do not hire undocumented migrants; and, of course, the dogmatic differentiation between "undocumented migrants", called "clandestini" and seen as criminals, and "honest labourers". The "amnesty", Salvini contends, is a useless and hazardous measure, because it will become a pull factor for irregular migration (the landings of migrants on the Italian shores – he claims – were four times higher in 2020 than in 2019, precisely thanks to the promise of regularisation) and, most of all, because it will "give a document for six months to those who now are probably dealing drugs in front of the main station in Rome and Milan" (Document 9).

In his usual populist speech style, during an intervention at the Italian Senate, to illustrate his "Italians first" slogan, he tells the history of a farmer, who wants to hire only Italian labourers:

> A farmer from the province of Cuneo [in the Piedmont region], I don't know him [...] Michele Ponzo, that owns a farm, he does not use the amnesty [...] this serious farmer, from Saluzzo, says: "in 48 hours I received 400 requests to come and work in the fields mostly from Italian citizens. This year a cook, a graphic designer and a 60-years-old Italian hairdresser will come and work with me." Before crying for the poor *clandestini*, please, take care of jobless and hopeless Italians!
>
> (Document 9)

This Manichean differentiation between honest (migrant) workers and "clandestini" was evident during a television debate with Aboubakar Soumahoro, the third actor of this public debate. When Soumahoro, as the leader of a grassroots union, announced a strike against the – in his opinion – too harsh requirements of the regularisation programme, Salvini, laughing, asked: "can the *clandestini* strike now? In which country do we live?", thus arguing that undocumented migrants simply *cannot* be workers (Document 4).

On his part, Aboubakar Soumahoro criticises the minister Bellanova precisely for her "utilitarian approach" to migration. In his view, in the context of an epidemic, undocumented migrants should be granted a permit of stay not because they are helpful for the Italian economy, but because their lives are at risk. As we have seen, this "humanitarian" concern is shared by Bellanova. Nonetheless, Soumahoro notes that only those who worked as farmworkers or domestic workers are allowed to request a permit of stay. In a radio interview, he claims that the measure should be applied to all "human lives", not only to few categories of workers:

> The strike of the invisible has been launched in a context of pandemic like the current one, where more than 32,000 human lives have been stolen from our community by this invisible enemy. [The government] is worried about the farm workers, the invisible and the fruit and vegetables, but it didn't worry about the safeguarding of persons' lives, differently from our doctors, who are saving human lives. The government decided to pay attention to fruit and vegetables that risk to rot, more

than to release to everybody a permit of stay for the health emergency that could be converted to a permit for labour reasons.

(Document 8)

In a television debate with Matteo Salvini, quoting Pope Francis, Soumahoro specifies:

> The State, the government, must urgently approve a measure to save lives first of all. Because those persons are not a consumer good, a disposable good, how Pope Francis would say [...] You should not regularise for market reasons, you should not regularise to avoid fruit and vegetables to rotting; you should regularise to save lives. Their lives are as valuable as mine and yours, Senator Salvini.

(Document 4)

Soumahoro's argument is the same as that which was used by Bellanova: undocumented migrants must be regularised for humanitarian reasons, because of the health risks due to Covid-19. They both use the same term to name this category of workers: the "invisible" ("*gli invisibili*"). As we have seen, Bellanova claimed that "From today, thanks to this Government's decision, the invisible will be less invisible" (Document 5).

Of course, the "humanitarian" moment is differently articulated in Soumahoro's discourse than in that of the minister. The union leader, indeed, accuses the "food giants" – mainly the big retail chains – of being mainly responsible for the poor conditions not only of migrant farmworkers, but also of Italian "peasants and farmers":

> I am here in the mud of poverty, where peasants and farmers are squeezed by the excessive power of the big retailers [...] an excessive power that squeezes peasants and farmers, who, in turn, squeeze the farmworkers.

(Document 4)

While organising the "strike of the invisible" on May 21, Soumahoro and his union called for the support of the Italian critical consumers, who, through a "shopping boycott" should solidarise with the farm labourers and ask for better work conditions in food production:

> For these reasons, as well as for the issue of exploitation within the agri-food chain, which is due to the concentration of power in the hands of the food giants, we decided to launch this strike, which will concern not only the farm labourers, but will receive the solidarity from consumers who will do a shopping boycott.

(Document 8; see also Document 7)

If Salvini's political discourse is clearly a form of radical right-wing populism – with its features of nativism and authoritarianism (Mudde, 2007;

Iocco et al., 2020) – Soumahoro's position can be considered a form of left-wing populism (see Stavrakakis & Katsambekis, 2014; Mudde & Rovira-Kaltwasser, 2018). He calls for a cross-class coalition, which should include other categories of migrant workers and Italian farmworkers, but also members of other social classes, such as farmers and peasants, and middle-class urban consumers, who – from a non-populist Marxist point of view – could be considered as bearing different class interests:

> A great number of workers are participating in the assemblies that are taking place. Thousands of female and male consumers are expressing their solidarity, through a shopping boycott. [...] Many people want to participate and are determined to put the regularisation of everybody at the fore, from the riders to the construction workers and the peddlers, in every sector hit by the health crisis that we are still living through. It isn't only a demonstration for farm labourers, but an initiative that tries to unite the hopes and sufferings of all the invisible and all of the people that are at the margins of the society.
>
> (Document 8)

The second element of populism in Soumahoro's discourse is common to all three political actors I analysed in this article, namely the claim to belong to the "people" and to "get their boots dirty" while walking in the muddy fields. In the television debate with Salvini, Soumahoro challenges him to "put his boots on" and "come to the fields together with us to listen to the peasants": "Come to the fields and let's discuss about the *caporalato*, that we fight by staying here on the fields, not in the rooms" (Document 4).

Salvini, concordant with his populist image of a political leader who knows his "people" well, promptly answers:

> Look, I probably visited a greater number of farms than you did [...] I am usually accused of going to too many farms instead of staying in my office. I know so many farms....
>
> (Document 4)

In the same way we can read Teresa Bellanova's history of her childhood as an exploited farmworker, which she could not forget when she entered the Ministry's rooms. In sum, all these political actors appear engaged in preventing and dismissing accusations of being part of the "political elite".

5.4 Populism as method: concluding remarks

This chapter has analysed the political debate on the "regularisation" of undocumented migrants, which was launched by the Italian government in the months of April and May 2020, during the Covid-19 health crisis. This measure concerned those "irregular" migrants who either were working or had worked in agriculture and domestic labour. In both cases, the government deemed undocumented workers as "deserving" of a permit of residence, not only because they were "necessary" to these key sectors of the Italian economy, but also because they "heroically" contributed to feeding the Italian people and to taking care of elderly Italian people in an extremely risky situation. The political debate mainly concerned the situation of undocumented migrants employed in agriculture. Through the analysis of the political discourses of three key actors in this debate – the Minister of Agriculture Teresa Bellanova, the leader of the centre-right coalition Matteo Salvini and the grassroots union organiser Aboubakar Soumahoro – the conclusions of this article are twofold.

First, all three actors incorporated traits of populism in their political discourse, although these traits are differently articulated (Laclau & Mouffe, 1985) with other elements. Both Bellanova and Salvini can be defined as "gastronationalists": they take for granted the "quality" and "excellence" of Italian food and agriculture, in the traditionally hegemonic discourse on *Italian* agriculture (Brunori et al., 2013; Iocco et al., 2020). However, this trait is differently articulated when it comes to farm labour: while Bellanova connects "gastronationalism" with a utilitarian and humanitarian approach to migrant labour, Salvini turns the "Made in Italy consensus" into both a nativist ("Italians first") and an authoritarian (he proposes an iron fist against undocumented migrants, seen as "clandestini") discourse, a typical expression of radical-right populism (Mudde, 2007). Moreover, their discourses include another populist trait: the idea that "the cause of the troubles is ultimately never the system as such but the intruder who corrupted it" (Žižek, 2006, p. 555), such intruder being identified with the illegal immigrant by Salvini, and with the *caporalato* and the mafia by Bellanova. Differently, Soumahoro's populism is a form of radical-left populism (Mudde & Rovira-Kaltwasser, 2018): he shares Bellanova's "humanitarian approach" but challenges her utilitarianism and calls for an alliance between different social classes (migrant and Italian farmworkers, other migrant labourers, peasant and farmers, middle-class urban consumers) against the economic elite, namely the "big companies", especially the retail chains, deemed responsible for the poor working conditions of peasants and migrant workers.

The second point concerns what I propose to call "populism as method": these three actors seem to ultimately legitimate their political discourses by representing themselves as *members of the people*, leaving aside almost all "technical" arguments in support of their claims. Soumahoro and Salvini compete on the basis of "getting their boots dirty" (i.e. how many farms they have visited). Bellanova recalls her childhood as an exploited farmworker. In this sense, the Italian political debate on agriculture and migrant farm labour is characterised by populism without any reference to a scientific technocracy that supports the political measures and claims, as happens in other forms of "post-politics" or "post-democracy" (Mouffe, 2005). Only as one of the people

– the actors of this debate seem to claim – can one legitimately discuss Italian agriculture: any other argument is of secondary importance.

Acknowledgements

The author wishes to thank Giulio Iocco and the editors of this volume for their suggestions and revisions on the first draft of this paper.

Appendix: list of documents cited

1. Teresa Bellanova, speech to the Italian Senate, Rome, 16 April 2020, www.youtube.com/watch?v=JDMIsvft8Zw. (last accessed: 24 September 2020).
2. Matteo Salvini, press conference at the Italian parliament (with Gianmarco Centinaio), Rome, 16 April 2020, www.facebook.com/salvi-niofficial/videos/260262808478676/?__so__=channel_tab&__rv__=all_videos_card (last accessed: 24 September 2020).
3. Teresa Bellanova, Interview to the newspaper *Corriere della Sera*, 8 May 2020, www.youtube.com/watch?v=l4anI8_XNg4 (last accessed: 24 September 2020).
4. Matteo Salvini and Aboubakar Soumahoro, Debate during the tv talk show "Mezz'ora in più" (Rai Tre), 10 May 2020, https://fr-fr.facebook.com/252306033154/videos/231699531427265/?__so__=watchlist&__rv__=video_home_www_playlist_video_list (last accessed: 24 September 2020).
5. Teresa Bellanova, Press Conference of the Italian Government, for the Presentation of the Law Decree "Decreto Rilancio", 13 May 2020, www.youtube.com/watch?v=rjyAOtJ8Enw (last accessed: 24 September 2020).
6. Teresa Bellanova, "You can't forget the exploitation", Interview to the news report TGCom24, 14 May 2020, www.facebook.com/310949775693438/videos/659441751575378/?__so__=channel_tab&__rv__=all_videos_card (last accessed: 24 September 2020).
7. USB Lavoro Agricolo, "Decreto Rilancio, sciopero di lavoratori della terra e invisibili: giovedì 21 cesti di frutta e verdura alle prefetture. Appello ai cittadini per uno sciopero della spesa", Press Release, 18 May 2020, www.usb.it/leggi-notizia/decreto-rilancio-sciopero-di-lavoratori-della-terra-e-invisibili-giovedi-21-cesti-di-frutta-e-verdura-alle-prefet-ture-appello-ai-cittadini-per-uno-sciopero-della-spesa-1623.html (last accessed: 24 September 2020).
8. Aboubakar Soumahoro, "Sciopero degli invisibili" ("Strike of the Invisible ones"), Interview at the Radio News Radio Popolare, 21 May 2020, www.radiopopolare.it/sciopero-degli-invisibili-21-maggio-intervista-a-aboubakar-soumahoro/ (last accessed: 24 September 2020).
9. Matteo Salvini, speech at the Italian Senate, 21 May 2020, www.youtube.com/watch?v=8L2b0vZIDbo (last accessed: 24 September 2020).

Notes

1   The Law-Decree 34/2020 was approved by the government on 19 May and by the parliament on 17 July (Law 77/2020). Article 103 is devoted to the regularization program.

2   "La Repubblica", 28 February 2020, www.repubblica.it/cronaca/2020/02/28/news/coronavirus_braccianti_stranieri_in_fuga_raccolti_a_rischio-249819390/ (last accessed: 24 September 2020).

3   "Il Corriere della Sera", 20 March 2020, www.corriere.it/economia/aziende/20_marzo_30/grano-pomodori-frutta-campi-servono-200mila-operai-604f16c2-72c3-11ea-bc49-338bb9c7b205.shtml (last accessed: 24 September 2020).

4   See, for example, the case of a vineyard grower in the Trentino-Alto Adige region, who rented a private flight for "his" eight Romanian seasonal women workers (ANSA Press Agency, 19 May 2020, www.ansa.it/trentino/notizie/2020/05/19/un-jet-privato-per-le-lavoratrici-stagionali-dalla-romania_798d4bfd-9558-446b-9163-08581ec3af52.html, last accessed: 24 September 2020).

5   With the aim of estimating the presence of "irregular" migrants in Italy, we can refer to the data on asylum seekers whose asylum request was denied in recent years. Indeed, since 2011, the majority of "new" migrants have been entering Italy as asylum seekers, mainly landing on its Southern shores, a process that has been defined as the "crisis of the European border regime" (New Keywords Collective, 2016; Campesi, 2018). Between 2014 and 2019, about 500,000 foreign citizens applied for international protection in Italy. Among them, around 300,000 were denied, but very few have been repatriated. They probably remained in Italy without any permit of stay, even though some could have moved to other European countries. Over the same period, about 85,000 asylum seekers received a permit of stay for "humanitarian protection" (for these data, see the website of the Minister of the Interior, www.libertaciviliimmigrazione.dlci.interno.gov.it/it/documentazione/statistica/cruscotto-statistico-giornaliero). In 2019, the "security decrees" approved by the Italian government when Matteo Salvini served as Minister of the Interior abolished this kind of residence permit; the migrants that did not apply for a residence permit for labour reasons before the expiration of the previous permit of stay for "humanitarian protection" become "irregular" as well. It is thus possible to estimate the presence of 300,000–400,000 undocumented workers in Italy. The Minister Bellanova proposed an estimate of 600,000 irregular migrants.

6   It should be noted that the measures of "amnesty" of undocumented migrants have been a recurrent instrument of Italian migration policies. Eight "amnesties" were launched from 1982 to 2012, from both centre-left and centre-right governments. In 2002, in the framework of the so-called "Bossi-Fini Law" (Law 189/2002), approved by the government headed by Silvio Berlusconi, about 640,000 undocumented migrants received a permit of stay; from 1986 to 2002, approximately 1.4 million migrants obtained their residence permit through a regularization program (Finotelli & Sciortino, 2009).

7   The deadline for the submission of the regularization request was 15 August 2020; at this date, about 207,000 demands were presented; 15% of them were submitted by agricultural labourers.

8   In this approach, "discourse analysis" is viewed as the "practice of analysing empirical raw materials and information as discursive forms [...] discourse analysts treat a wide range of linguistic and non-linguistic data – speeches, reports,

manifestos, historical events, interviews, policies, ideas, even organisations and institutions – as 'texts' or 'writing'" (Howarth & Stavrakakis, 2000, p. 4).

9 "Moments" are defined by Laclau and Mouffe as "differential positions" that "appear articulated within a discourse", while "articulation" is defined as "any practice establishing a relation among elements such that their identity is modified as a result of the articulatory practice" (1985, p. 105).

10 Until 2013, the party was named "Lega Nord" (Northern League), used to win its votes almost exclusively in the Northern regions and was characterized by an anti-Southern disposition. From 2013 on, new leader Matteo Salvini gave the party a nationalist turn, thus becoming the most important party of the centre-right coalition and, at the European elections of May 2019, the most voted-for Italian party, with 34.1% of the votes.

11 According to the estimates of the Osservatorio Placido Rizzotto (2012), a research centre connected to the Flai-CGIL the main labour union in the agri-food sector, about 400,000 workers (both migrants and Italians) are "at risk" of being severely exploited by *caporali* and employers. This estimate has been widely used in the political debate on this issue over the last years.

References

Avallone, G. (2016). "The land of informal intermediation: The social regulation of migrant agricultural labour in the Piana del Sele, Italy", in Corrado, A., de Castro, C. and Perrotta, D. (eds) *Migration and agriculture: Mobility and change in the Mediterranean Area*. London: Trade-off or convergence? The role of food security in the evolution of food discourse in ItalyRoutledge, pp. 217–230.

Benasso, S. and Stagi, L. (2019). "The carbonara-gate: Food porn and gastro-nationalism", in Sassatelli, R. (ed.) *Italians and food*. London: Palgrave McMillan, pp. 237–267.

Brunori, G., Malandrin, V. and Rossi, A. (2013). "", *Journal of Rural Studies*, 29, pp. 19–29.

Campesi, G. (2018). "Crisis, migration, and consolidation of the EU border control regime", *International Journal of Migration and Border Studies*, 4 (3), pp. 196–221.

Corrado, A., De Castro, C. and Perrotta, D. (2016). "Cheap food, cheap labour, high profits: Agriculture and mobility in the Mediterranean. Introduction", in Corrado, A., de Castro, C. and Perrotta, D. (eds) *Migration and agriculture. Mobility and change in the Mediterranean area*. London: Routledge, pp. 1–24.

Corrado, A., Lo Cascio, M. and Perrotta, D. (2018). "Introduzione. Per un'analisi critica delle filiere e dei sistemi agro-alimentari in Italia", *Meridiana. Rivista di storia e scienze sociali*, 93, pp. 9–29.

Crea [Consiglio per la ricerca in agricoltura e l'analisi dell'economia agraria] (2019). *Il contributo dei lavoratori stranieri all'agricoltura italiana*, ed. by M. C. Macrì, online: www.crea.gov.it/web/politiche-e-bioeconomia/-/on-line-il-contributo-dei-lavoratori-stranieri-all-agricoltura-italiana (last accessed: 24 September 2020)

DeSoucey, M. (2010). "Gastronationalism: Food traditions and authenticity politics in the European Union", *American Sociological Review*, 75 (3), pp. 432–435.

Finotelli, C. and Sciortino, G. (2009). "The importance of being Southern. The making of policies of immigration control in Italy", *European Journal of Migration and Law*, 11, pp. 119–139.

Fonte, M. and Cucco, I. (2015). "The political economy of alternative agriculture in Italy", in Bonanno, A. and Busch, L. (eds) *Handbook of the international political economy of agriculture and food*. Cheltenham: Edward Elgar, pp. 264–294.

Grasseni, C. (2013). *Beyond alternative food networks. Italy's solidarity purchase groups*. London: Bloomsbury.

Howard, N. and Forin, R. (2019). "Migrant workers, 'modern slavery' and the politics of representation in Italian tomato production", *Economy and Society*, 48 (4), pp. 579–601.

Howarth, D. and Stavrakakis, Y. (2000). "Introducing discourse theory and political analysis", in Howarth, D., Norval, A. J. and Stavrakakis, Y. (eds) *Discourse theory and political analysis. Identities, hegemonies and social change*. Manchester: Manchester University Press, pp. 1–23.

Iocco, G., Lo Cascio, M. and Perrotta, D. (2020). "'Close the ports to African migrants and Asian rice!': The politics of agriculture and migration and the rise of a 'new' right-wing populism in Italy", *Sociologia Ruralis*, Online First, First published: 20 April 2020, 60, pp. 732–753.

Laclau, E. and Mouffe, C. (1985). *Hegemony and socialist strategy. Towards a radical democratic politics*. London: Verso.

Mouffe, C. (2005). *On the political*. London: Routledge.

Mudde, C. (2007). *Populist radical right parties in Europe*. Cambridge: Cambridge University Press.

Mudde, C. and Rovira-Kaltwasser, C. R. (2017). *Populism: A very short introduction*. Oxford: Oxford University Press.

Mudde, C. and Rovira-Kaltwasser, C. R. (2018). "Studying populism in comparative perspective: Reflections on the contemporary and future research agenda", *Comparative Political Studies*, 51 (13), pp. 1667–1693.

New Keywords Collective (2016). *Europe/crisis: New keywords of "the crisis" in and of "Europe"*. Online at: http://nearfuturesonline.org/europecrisis-new-keywords-of-crisis-in-and-of-europe/ (last accessed: 24 September 2020).

Osservatorio Placido Rizzotto (2012). *Agromafie e caporalato. Primo rapporto*. Rome: Flai-CGIL.

Palumbo, L. and Corrado, A. (eds) (2020). Covid-19, Agri-food systems and migrant labour. The situation in Germany, Italy, The Netherlands, Spain, and Sweden, Report, July 2020, European University Institute and Open Society Foundations. Online at: www.opensocietyfoundations.org/uploads/ccf241cc-89b2-4b44-a595-90bd77edab3d/covid19-agrifood-systems-and-migrant-labour-20200715.pdf (last accessed: 24 September 2020).

Perrotta, D. and Sacchetto, D. (2014). "Migrant farmworkers in Southern Italy: Ghettoes, *Caporalato* and collective action", *International Journal of Strikes and Social Conflicts*, 1 (5), pp. 75–98.

Potter, C. and Tilzey, M. (2005). "Agricultural policy discourses in the European post-Fordist transition: Neoliberalism, neomercantilism and multifunctionality", *Progress in Human Geography*, 29 (5), pp. 581–600.

Salvia, L. (2020). "The restructuring of Italian agriculture and its impact upon capital–labour relations: Labour contracting and exploitation in the fresh fruit and vegetable supply chain of the Lazio Region, Central Italy", *Journal of Agrarian Change*, 20 (1), pp. 1–15.

Stavrakakis, Y. and Katsambekis, G. (2014). "Left-wing populism in the European periphery: The case of Syriza", *Journal of Political Ideologies*, 19 (2), pp. 119–142.

Žižek, S. (2006). "Against the populist temptation", *Critical Inquiry*, 32, pp. 551–574.

6 Political and apolitical dimensions of Russian rural development

Populism "from above" and *narodnik* small deeds "from below"

Alexander Nikulin and Irina Trotsuk

Introduction

Populism is both a well-known and an extremely vague, slippery concept (Panizza, 2005: 1). It is variously defined as an ideology, a strategy, a discourse, or a political logic (Moffitt, 2016: 5). It appeals to real or imagined voting majorities ("people" as an imagined homogeneous body sharing interests and believing in a leader, whose mission is to save the nation) to achieve political, institutional power and use authoritarian ways to wield power (Roman-Alcalá, 2020: 16). In political analysis, already in 1969 Ionescu and Gellner (1969) could not settle on a unitary definition of populism, and half a century later it is still a contested concept. As Brubaker rightly notes,

> few categories in the social science lexicon have been more heatedly contested in recent years than "populism". The conceptual meaning (a discursive form or style), empirical extension (substantive political commitments or social structural foundations) and normative valence (endemic in democratic settings or episodic, dangerous or desirable) of the category are all deeply disputed ... Populism has been stretched so far in some scholarly usage – and in much journalistic discussion and public commentary – that it risks dissolving as a distinct object of analysis.
>
> (Brubaker, 2020: 44–45, 53)

However, there is scholarly agreement on the analytical core of populism – as "an anti-phenomenon and as a people-worshipping phenomenon" (an idealised conception of the people's community in opposition to the status quo that is stealing power from the "people") (Deiwiks, 2009: 1). Moreover, there are two general approaches to the study of populism: theoretical (historical and contemporary), and the analysis of specific manifestations of populism across countries and regions. Within the theoretical approach, Brubaker (2020) identifies two tightly interwoven, even mutually constitutive oppositions that are also useful for the study of regional types of populism: vertical opposition to "those on top and those on the bottom", and horizontal opposition to outside groups. Thus, economic, political and cultural elites are represented as both "outside" and "on top" in all populist discourses.

DOI: 10.4324/9781003091714-7

The majority of works on populism focus on its specific cases (including Russia) rather than on theoretical discussions about the essence of the concept (Taggart, 2000). Due to limitations of space, we omit a review of the general debates about definitions, changes and types of contemporary populism to focus on the current, populist trends in Russia, which makes this text represent the second approach – regional-specific. In this chapter, we identify two types of Russian populism: the official, politicised populism "from above" (mainly authoritarian) and a specific Russian form of populism "from below", which is historically rooted in the *narodnik* tradition and in the "theory of small deeds". Both are based on a distinction between the people and the elites, but unlike populism "from above", populism "from below" is apolitical, decentralised, cultural and considers social goals more important than political ones.

This chapter aims to identify the "canonic" and specific features of populism[1] as represented in contemporary Russian realities. The analysis is based on the theoretically and empirically studied differences between the political, discursive approaches to rural development issues and the everyday practices of enthusiasts who strive to save the disappearing village as a special lifestyle. We do not question the "type" of Putin's populism, but possible Russian populisms: their origins, features and prospects for future development. Therefore, the chapter contributes to the global debates on the agrarian question (the economic and political consequences of capitalist relations for the peasant economy; see, e.g., Bernstein, 2004) and on 'persistence versus disappearance' (see, e.g., Edelman, 2005; Wegren, 2006; Scoones, Edelman, Hall, Wolford, & White, 2018; Ivanou, 2019). In rural Russia, persistence stands for the moral economy that strives to survive under authoritarian populism "from above". This has led to the disappearance of traditional, rural life due to the state support of huge agro-holdings, which oust other forms of rural economies. The key driver for the survival of this moral economy[2] is a kind of revived *narodnik* movement – "from below" – a very special type of Russian populism.

In Russia, the term "populism" is ambiguous, and confusion starts with translating the Russian term "*narodnik* movement" as an English word – "populism".[3] Was the Russian *narodnik*, a powerful social, ideological movement in the late 19th and early 20th century, a kind of "standard" populism? According to the "standard" definition of populism of that time, it was a political movement in search of a strong, charismatic leader capable of releasing the "living creativity of the masses" in the fight against the inertia of the state bureaucracy and the omnipotence of the capital. This definition implied the revival of national values as preserved mainly in the Russian peasantry and the Russian village. However, only in its most marginally conservative part was the *narodnik* movement a synonym for populism. *Narodnichestvo* was mainly a left-liberal ideology of the peasant enlightenment with the help of the developing rural institutions of local self-government, cooperation and education. There were some features of traditional authoritarian (both right and left) populism in the party ideologists – from

Monarchists to Social Revolutionaries and Bolsheviks. However, the three "isms" of Monarchism, Social Revolutionarism and Bolshevism are much broader and more complex than "standard" populism (see, e.g., Tikhonova, 2011; Solovieva et al. 2014; Fedyunin & Pain, 2019).

A convincing explanation of the ideological difference between the Russian terms "populism" and *"narodnichestvo"* was provided by Furman:

> If simplified, populism is still a political technology that does not imply a radical transformation of the society; while *narodnichestvo* tries to change the society in such a way that there are no "simple" people about whom someone should take care of.
>
> (Furman, 2009: 113)

Thus, the populist leader strives constantly to (re)establish the rules of the game between the elites and the people (theoretically) in favour of the people, whereas the *narodnik* aims to create a democratic society in which the fatal division into "sophisticated" elites and "rude" people is eliminated.

In the 1990s, during the national economic crisis, there was a so-called "red belt" on the Russian electoral map. This consisted of regions that always voted against Yeltsin's policies and liberal reforms, demanded the return of the Soviet welfare state and truly supported the conservative values of nationalism and traditionalism promoted by the Communist Party and the Liberal-Democratic Party. This "red belt" was mainly a "rural belt" (Lindner & Nikulin, 2004); it consisted of agrarian regions with the highest share of rural population. Some liberal critics argued that the highest protest vote came from this "belt" and was typical for the central, Black-Earth area of European Russia – that is, the regions of the historically most sustainable and cruel peasant serfdom (see, e.g.: Davydov, 2018). Liberal critics claimed that the notorious tradition of Russian serfdom was manifested in the yearning of this poor, rural population for a strong, national leader who would put an end to the speculations of liberal financiers and protect the patrimonial, collective farms from inevitable bankruptcy under the free market of the 1990s (see, e.g., Amelina, 2000).

In Putin's era, liberal critics discovered that by 2010, the electoral, rural "red belt" had evolved into "Putin's belt" (Wegren & Konitzer, 2006). In the early 21st century, rural Russia, the country of villages and small towns, became a true stronghold of Putin and his party, United Russia (Artemiev, 2010). The geographer-economist Zubarevich (2011) developed a theory of "four Russias": the "first Russia" is developed, liberal and concentrated in two capital cities and the other largest cities of the country; the "fourth Russia" is a peripheral domain of rural unemployment, poverty, poor roads and mass alcoholism. Between them are two moderate types – the "second Russia" and the "third Russia". The "first Russia" is a country of liberal opposition to Putin's regime, and the "fourth Russia" is a country of pro-Putin conservative populism. This scale is supported by many political researchers (see, e.g., Petrov et al., 2013; Granberg and Sätre, 2016; Rodoman, 2017), but the question is whether Putin corresponds to the image of the authoritarian-populist politician who relies mainly on the traditional rural values and electorate. What are the other types of authoritarian politicians that represent rural populism in Russia?

Populism "from above": Putin and typical, regional, populist leaders

When Putin came to power, Russian agriculture was in an extremely difficult situation. In terms of the production of basic agricultural products, Russia was thrown back several decades to the level of the 1960s. Under the free market system, the majority of state and collective farms were in their death throes because of their inability to become effective, commercial enterprises. The new stratum of farmers was small, its economic successes were modest and the market and state bureaucratic inertia hindered its sustainable development. Unemployment, low incomes, absenteeism and alcoholism were key features of rural life, and were aggravated by the increasing demographic depopulation. Agrarian development was therefore declared a priority of the economic policy (Uzun & Shagaida, 2015).

In the early 2000s, special national projects were initiated to support large agricultural producers, improve systems of rural education and health care, and provide rural areas with the Internet – mainly to strengthen the export position of agriculture and ensure national food security (Trotsuk, Nikulin, & Wegren, 2018). Today, the agrarian rhetoric of the Russian president is rather authoritarian and autarchic: he frequently mentions the need to take care of the rural population, increase rural incomes and develop and improve the rural social sphere, and emphasises that the agricultural producer must ensure autarkic, national food security (Wegren, Nikulin, & Trotsuk, 2019). This is a typical feature of populism – that is, the invocation of nationalism in opposition to internationalism and externally oriented capital:

> For populism, then, it is not capitalism per se that is the problem, but rather "foreign", "big", "corporate", or transnational capital. It is not difficult to see, then, how populism overlaps with national, "progressive", or indeed "populist" understandings of food sovereignty.
>
> (see, e.g., Roman-Alcalá, 2020)

The emphasis is placed on the national "local" as opposed to the international "corporate" (McMichael, 2013). Certainly, Putin's agrarian ideology has some elements of rural, authoritarian populism, especially in calls for national independence in food security. However, this ideology is too bureaucratised: Putin rarely appeals to the rural population directly; he prefers to communicate with the local and regional bureaucracy responsible for rural development and large agribusiness. Putin consistently spreads his authoritarian, populist ideology in the countryside, as is reflected in the fact that it was under his rule that rural municipalities lost any independence and became subordinate to the state bureaucracy (see, e.g., Berelowitch, 2011; Latukhina, 2019).

There are various types of rural, authoritarian populism among regional leaders. They can be demonstrated with reference to three local figures: Yevgeny Savchenko, Pavel Grudinin and Vasili Melnichenko. The governor of the Belgorod Region, Savchenko[4] is one of the "governors-long-livers" (he has held his office since 1993, almost without interruption), who managed to make his region a demonstration model of successful agricultural production and rural development (Nikulin, Trotsuk, & Wegren, 2017; Wegren, Nikulin, & Trotsuk, 2018). Like Putin, Savchenko is a consistent etatist, but he is a much stronger critic of the liberal economy. He combines the right and left populist ideas of Stolypin and Chayanov with ideals of the Russian Orthodox Church and classic literature, and claims to be an arbiter in the search for compromises between large agribusiness and regional, environmentally friendly, family and community production (Savchenko, 2020).

In contrast, the head of the agricultural enterprise, "Lenin State Farm", in the Moscow Region, Pavel Grudinin is an example of agrarian-populist ideology criticising contemporary Russian etatism and bureaucracy for supporting large agribusiness. Grudinin is a talented head of the enterprise that produces strawberries and milk on 2000 ha and a star of political talk shows about the key challenges of rural Russia. In 2018, in the last presidential elections, Grudnin was an independent candidate (supported by the Communist Party) and achieved the impressive result of second place after Putin (Ivanov, 2019). In his election campaign, he criticised Putin for his bureaucratic leadership style and for state support for large agribusiness, which has led to an increase in social and regional inequality, stagnation and corruption. It was Grudinin's typical habit to address the masses directly and emphasise the successes of his social programmes. By describing his enterprise's advantages, Grudinin wanted to prove the possibility of the country's return to Soviet, nationwide, social achievements provided by the elimination of bureaucratic, oligarchic, arbitrary rule. This anti-bureaucratic rhetoric against the fundamental elements of national rural policy and Putin's state provoked a repressive response against Grudinin's enterprise. As a result, he was accused of economic violations. There have been attempts to raid and grab his enterprise (according to Edelman, Oya, & Borras, 2013)[5] and discredit his work as an entrepreneur and politician (Titov, 2020).

Melnichenko believes that anti-Putin protests are unpromising because their participants know nothing about real life:

> I am sure that today the only person who can change the situation in Russia is Vladimir Putin... The State Duma is a bunch of talkers; the government is a bunch of thieves... The main thing is that the President should have fewer advisers with legal education, because not a single document compiled by these "know-alls" can be understood. And there should be fewer economists by the President, because they think of themselves as "experts", but if you ask them a simple question of what manure is better – swine or cow – they would stress that they are "experts" and would not answer the question... One must first learn to distinguish

varieties of manure, and only then come and teach peasants... I do not criticise political protests for they are a surge of civil activity and a normal reaction to the surrounding absurdity; however, protesters are still wet behind the ears and unfamiliar with the work on land.

(Bozhenko, 2013)

To conclude this brief review of typical, populist ideologies of contemporary Russia, it is necessary to emphasise their paradoxical inconsistency. Thus, Putin aims at the sustainable development of rural Russia by supporting large agribusinesses and regional bureaucracies, which are the main accelerators of the economic polarisation and administrative centralisation of rural areas, which exacerbates the general degradation of the countryside. Rural reformers, such as Savchenko (at the regional level) and Grudinin (at the local level), call for the democratic coexistence of large, medium-sized and small agricultural enterprises and for the national expansion of social development programmes. However, they insist that such coexistence and expansion should be ensured by an authoritarian state of the Soviet, socialist type (for instance, both Savchenko and Grudinin argue that Stalin was too cruel, but pursued the right policy). Melnichenko calls for freedom for the rural population and believes that this freedom can be granted only by the president (Shatskikh, 2016). Since *perestroika*, Melnichenko has founded many small agricultural enterprises but has not achieved any success, unlike Savchenko, who made the Belgorod Region a leader of regional, social-economic development, and Grudinin, who created a unique, post-state farm. Savchenko does not criticise Putin – he questions the liberal course of his ministers; Grudin reproaches Putin for his economic policy with no prospects. Both Savchenko and Grudinin developed balanced programmes for national social-economic development. Melnichenko never criticises Putin and has not yet presented an economic programme. He is an active supporter of Putin and an uncompromising critic of his ministers and experts.

In terms of populism, Savchenko, Grudinin and Melnichenko pursue a typical authoritarian-populist idea of being a "supreme arbiter" in relations between the elite and the people and of acting in the interests of the people. Such an intent is consistent with the conventional definition of populism as "a movement towards a dominating and 'authoritarian' form of democratic class politics – paradoxically, apparently rooted in the 'transformism' (Gramsci's term) of populist discontents" (Hall, 1985: 118), and as "the modalities [sometimes considered demagoguery] of political and ideological relationships between the ruling bloc, the state and the dominated classes" (Hall, 1985: 119). The differences among the three Russian populists are determined by the paths they suggest and follow within the authoritarian-populist strategy of national rural development. On the one hand, these differences are implied in their political and economic power: Savchenko[6] (like Putin) is a representative of the ruling power, whereas the heads of agricultural enterprises – Grudinin and Melnichenko – are (allegedly) representatives of the opposition. On the other hand, these differences are determined by ideological priorities within the contradictory field of authoritarian populism: Putin supports political authoritarian conservatism and economic liberalism, whereas Savchenko and Grudinin prefer a form of authoritarian-conservative socialism. (Melnichenko does not have a clear ideology and only supports a strong leader (Putin) who must listen to the people's wisdom.)

Should we consider contemporary Russian populism as only populist authoritarian ideas of political figures, or are there also elements of the democratic populism of the *narodnik* type?

Populism "from below": rediscovery of the "theory of small deeds"

In pre-revolutionary Russia, there were two types of populism "from below": the revolutionary, radical *narodnik* movement and the moderate, reformist populism close to the "theory of small deeds". The Russian *narodniks* were a "classical" populist movement and Russia's main indigenous revolutionary tradition, which dated to the 1870s (Shanin, 1983). The "theory of small deeds" was a collection of the intelligentsia's social projects aimed at the gradual improvement of peasant life, public education and healthcare (Gordeeva, 2003). These projects contributed to the increase in the well-being and culture of the peasantry and to the development of *zemstvo* and cooperative movements in the 20th century. However, the moderate *narodniks* – the main supporters of the "theory of small deeds" – were reproached for their "petty political idealism" in the 1870s, and still are – that is, for being incapable of setting and achieving ambitious goals in the decisive struggle for the radical social and political reforms against inert statehood (Zverev, 2016).

Actually, there is no "theory" of small deeds in the scientific sense of the word – only some ideological, ethical ideas. For example, if it is not possible to do "great deeds" (to use a revolution to change the political system, or make fundamental social and economic transformations), one should focus on "microscopic" opportunities for changes for the better – participate in everyday local actions and develop education and healthcare for the poor and the peasantry. In the future, under a favourable situation for great deeds, the years and decades of small deeds would prove meaningful and fruitful, because the peasantry would be more educated and better prepared for socially responsible behaviour in a new era of radical social transformations (see, e.g., Korolenko, 2018/1965).

In the 21st century, there are ideas, projects and ideological disputes similar to the "theory of small deeds". Our field research and other sociological studies of rural Russia have discovered thousands of regional and local initiatives aimed at the gradual improvement of rural life by developing educational projects for the cultural, historical, recreational and ecological transformations of the countryside. There are hundreds of eco-villages and thousands of projects that have been implemented by activists. They define their ideological principles as rooted in or linked to the "theory of small deeds" – either in its humanistic ideas (such as the role of the intelligentsia in overcoming the gap between the "enlightened minority" and the "backward majority") or in its methods for improving rural life (such as developing social self-organisation and local self-government) (see, e.g., Zverev, 1997). The majority of such projects are not purely rural. Quite often, city dwellers move to the countryside and develop these projects; in other words, representatives of the "first Russia" create and support opportunities for the revival of the "fourth Russia". In general, the contemporary movement of "small deeds" stays out of active politics and pursues educational and environmental ideas and goals.

The contemporary Russian movement, which defines itself or is labelled by researchers as an heir to the "small deeds theory", is different from numerous grassroots initiatives all around Europe (and possibly all around the world), primarily because the Russian "small deeds" movement does not pursue political goals. Moreover, for reasons of political security, the new local forms of the *narodnik* movement are implemented as purely apolitical. Heads and representatives of the "small deeds" initiatives declare their absolute apolitical nature and insist that the local communities and authorities perceive them merely as activist, cultural and historical implementations of the wise federal policy of the Russian state. As a rule, in European grassroots initiatives, representatives of civil society aim to influence the political order and change it for the better. The Russian "small deeds" movement avoids the question of the political significance of "small deeds" and emphasises that such activities contribute to the improvement and humanisation of rural, provincial life. Thus, the movement does not oppose or contradict the aims of state social policies.

Certainly, there are similarities and differences between the "small deeds" of the early 20th and early 21st centuries. A century and a half ago, the activist forces of the Russian intelligentsia focused on the "people" – mainly the peasantry, which constituted the poorest and uneducated overwhelming majority of the population. In the early 21st century, the activist forces of the "theory of small deeds" lost its focus on the rural population, which has been increasingly outnumbered by the urban population (about 27%), and lost its traditional peasant base (Pivovarov, 2001). As before, the rural population is poorer and less educated than the urban population. However, today there are no comparable insurmountable, political, economic, class or cultural boundaries between rural and urban residents as there were between the peasantry and other social strata in tsarist Russia. The activist of "small deeds" recognises the ideological limitations of global consumer values and seeks new cultural, environmental ideals, which can be discovered, created, developed and enriched on the vast expanses of the Russian hinterland, which is often marginalised, depressed and abandoned, but preserves its unique potential for self-development (see, e.g., Magun, 2014).

Thus, the main concern of the activist of "small deeds" is some provincial space that has been abandoned and forgotten by the state and the market. Previously it was densely populated by the peasantry, but today is increasingly deserted and suffering from social emptiness and natural wildness. The activist can discover here the potential for revival, because of the unique composition of local, natural and cultural phenomena – landscapes, historical monuments, local crafts, natural resources, folk traditions, etc. This type of "small deeds" activism is not a true *narodnik* movement (see, e.g., Zverev, 1997), but rather a "spatial" movement aimed at the rediscovery and provision of necessary facilities for the local territory, so that it may become a new domain of comfort and culture for the future revival of a contented life (see, e.g., Kuznetsova, 2003).

Some ideas from Chayanov can serve as a theoretical and methodological bridge between the previous populist activity and contemporary "spatial" activists. On the one hand, Chayanov was a contemporary and a supporter of "small deeds". On the other hand, he was a futurist who predicted trends for both Russian and global rural–urban development (Chayanov, 2006). Moreover, Chayanov is considered a representative of the Russian peasant neopopulist tradition:

> The neopopulist tradition emphasised the viability of peasant agriculture and its ability to survive and prosper under any circumstances. For the peasantry had no necessary tendency to develop the increasing economic inequalities and class antagonisms of bourgeois industrial society; there was no tendency to create increasing groups of rich and poor or landless peasants with a more and more unstable group of middle peasants in between. The village was an overwhelmingly homogeneous community, able constantly to reproduce itself both economically and socially. Consequently, Chayanov saw the modernization of traditional small farming as lying along neither a capitalist nor a socialist road, but as a peasant path of raising the technical level of agricultural production through agricultural extension work and cooperative organisation, at the same time conserving the peasant institutional framework of the family small-holding.
>
> (Harrison, 1975: 390)

There are several paths for the development and implementation of the "theory of small deeds". The first and most widespread is the cultural approach that seeks to preserve and increase the growth points of local culture, which corresponds to the traditional *narodnik* "theory of small deeds". Based on this theory, in his utopias Chayanov persistently set for himself the task of decisively developing and integrating the local, rural, cultural heritage into country life (see, e.g., Nikulin, 2020). In the late 20th century, one of Chayanov's utopian reformer-heroes explained the cultural foundations for successful and sustainable rural urban development in Russia.

> We were afraid that our rural population scattered in forests and fields would gradually lose its vitality and culture... To fight such social "acidification", we needed a system of social "drainage"... We were plagued with the thought: are higher forms of culture possible with the scattered

rural population? We made every effort to build ideal transport routes. We found means to force population to move along these routes, at least to their local centres. We supplied these centres with all available cultural elements – *uyezd* and *volost* theatres, *uyezd* museums with *volost* branches, people's universities, sports of all kinds and forms, choir societies, i.e. everything, including the church and politics, was provided for the village to raise its cultural level. We risked a lot....

(Chayanov, 1989b: 211)

The experience of Soviet, Russian and international development of rural hinterlands proves that such a reliance on culture in the long-term perspective is an indispensable foundation for any sustainable regional development. In Russia, this approach is presented in the steady increase in the number of large and small projects in the cultural development of villages and small towns (Cultural mosaic, 2019).

The activists' striving for the diverse development of local cultures is accompanied by attempts rooted in the *narodnik* past to create sustainable and independent local communities. These range from establishing idealistic communities, such as eco-settlements, to pragmatic efforts to save local settlements from the destructive influence of centralising urbanisation, primarily from the outflow of the local population due to poor living conditions.[7] An example of such an attempt is the Ugric project[8] in the Manturov district of the Kostroma Region in the Russian North. A group of scholars from Moscow, who bought *dachas* (summer houses) in this rural hinterland started a systematic study of the social-cultural features of the territory and organised a series of research and scientific events on the prospects of "unpromising" villages in the Northern Non-Black-Earth regions (Pokrovsky & Nefedova, 2014). In addition to Chayanov's prediction of the significance of cultural activism, there are obvious echoes of his ideas about the importance of autarkic models of the natural economy for rural households and settlements (Chayanov, 1989a).

The remote autarkic communities in the northeastern regions of Russia have different trajectories of development, but some similar patterns. In the face of permanent crisis and depression, two types of hinterland settlements have been especially exposed to the risk of decreasing inhabitation (even depopulation). They are settlements located either in a busy trade place or in the middle of nowhere: the former are usually absorbed by the nearest cities to become their suburbs; the latter, godforsaken places disappear due to their fatal isolation. Only those remote autarkic communities that lose their spatial, geographic potential slowly are relatively sustainable, and are sometimes successful in social-economic development (Pozanenko, 2019). Eco-settlements are an extreme example of successful contemporary rural autarchy. Environmental activists strive to create almost autonomous households united in ecological communities that attract members with diverse religious-philosophical and esoteric-mystical ideals (Karpov, 2017).

Conclusion

There are two populist trends in contemporary Russia: the official one "from above", and the local one based on the "theory of small deeds". The latter is typical for the Russian North, but there are hundreds of economic, cultural and technological initiatives "from below" all around the country (Fadeeva, 2019; Frühauf, Guggenberger, Meinel, Theesfeld, & Lentz, 2020). However, such forms of local, rural self-organisation are not enough to improve the situation in rural territories. They should be supported at all levels of power, which would certainly correspond to the populist, political statements of federal and regional leaders. Otherwise, in the face of increasing administrative, economic centralisation, the implementation of the "theory of small deeds" related to culture, autarchy, a symbiotic combination of lifestyles, self-organisation and the multi-structural spatial development of local, rural territories will remain – in the old *narodnik* terms – "separate inspiring phenomena in the depressing social reality".

Both identified trends are definitely populist in Laclau's definition of populism as a kind of articulation of the antagonism between "ordinary people" and the authorities, which is implicit for the political field. Laclau considers the vagueness of this definition in the depoliticised environment of neoliberal capitalism as an advantage that allows the analysis of populism as an open movement beyond its historical, dogmatic logic, which can become a truly unifying and liberating movement for contemporary societies (see, e.g., Laclau, 2005). In Russia, populist political ideas failed to mobilise broad masses. This was due both to the neutralising efforts of the government and its claims to be the key "populist" "from above", as personified by the authoritarian leader Putin (hyper-politisation of populism, according to Panizza, 2005: 20; see the results in Mamonova, 2019). Russian populism "from below" turns out to be apolitical or depoliticised due to several reasons (in addition to the unwillingness and fear of dealing with the official conservative policy of the authoritarian state): first, disappointment in all official ideologies, movements and parties (liberal, conservative, communist) after the collapse of *perestroika* and the crisis of the 1990s; second, limited ability of too-local initiatives to effect political unification; third, in addition to spatial localism, the cultural-sectoral localism of "small initiatives" (for instance, activists of environmental projects are indifferent to local history or sports, and so on); fourth, the general level of social trust in Russia is still low, which also prevents the integration of local initiatives "from below" into a significant social-political movement. Therefore, the "theory of small deeds" follows the path of an apolitical improvement of rural life in such a way that can help the local population keep its moral economy and learn to oppose the capitalist market trend with cultural, educational and environmental initiatives that can be either independent from the state or skillfully embedded in its authoritarian, populist projects of rural development.

To conclude our short review of contemporary trends in Russian populism, it should be noted that various movements of "small deeds" in general adhere to liberal and tolerant views, avoid political life and do not have well-known ideological leaders on a national scale. This corresponds to the common definition of populism as not being only authoritarian. As Badiou (2016) emphasises, arguments in favour of "the people" can be a positive, mobilising force of solidarity, especially if the populist approach is combined with the moral economy concept, based on ideas of fighting capitalism openly or resisting it on a daily basis (Edelman, 2005).

There are ongoing debates about whether Putin's rule is authoritarian populism (many scholars agree that Putinism and populism have much in common; see, e.g., Berelowitch, 2017; Gudkov, 2017; Yudin and Matveev 2017; Edelman, 2020). Certainly, Putinism has features of political populism, such as a tendency to be simplistic and straightforward to appeal to the common sense of the ordinary people, and to propose transparent and easily understandable solutions to political problems (Deiwiks, 2009: 5). Putinism has features of authoritarian populism, such as a personal nature based on the figure of a charismatic individual who claims to be the incarnation of the people's will, spawns a "vulgarisation" of political discourse intended to shore up his anti-elitist and anti-cosmopolitan credentials, and tries to preserve at least a façade of electoral competition and legitimacy (Edelman, 2020). Moreover, Putinism is a part of the recent global rise of authoritarian populism, which

> typically depicts politics as a struggle between 'the people' and some combination of malevolent, racialised, and/or unfairly advantaged 'others', at home or abroad or both. It justifies interventions in the name of 'taking back control' in favor of 'the people', returning the nation to 'greatness'. 'Authoritarian populism' frequently circumvents, eviscerates, or captures democratic institutions, even as it uses them to legitimate its dominance and centralise power.
>
> (Scoones et al., 2018: 2–3)

In Russia today, the official, politicised, populist discourse "from above" does not appear to be practically effective and ideologically convincing, with its huge financial resources and "universal elements of populism" (Taggart, 2000: 54). On the other hand, the *narodnik* ideology of "small deeds", which emerged as a movement of Russian intellectuals in the 1870s to go from the cities to the countryside to "enlighten" the peasantry, appears to have been revived. Historically, it was essentially a populist movement because of its focus on the "people" and the distinction between the "people" and the "authorities". Yet, it was a rather anti-political movement that considered social goals more important than political ones (Canovan, 1981). The *narodnik* phenomenon "seems to be a highly unique case very different from populism elsewhere" (Deiwiks, 2009: 6), and its contemporary version is represented by the "theory of small deeds". The mainly apolitical, cultural, populist "small-deeds" movement "from below' really changes and improves rural life at the local level.

Notes

1 The chapter is based on the working definition of populism by Mudde: "An ideology that considers society to be ultimately separated into two homogeneous and antagonistic groups, "the pure people" versus "the corrupt elite", and which argues that politics should be an expression of the general will of the people" (Mudde, 2004: 543).

2 The term "moral economy" was propounded by Thompson and developed by Scott as an idea that social relations are grounded in a publicly recognised right to subsistence that entails reciprocal relations and obligations of social classes (the lower ones and the elites). Thompson (1971) combined a paternalist model of a society of consistent traditional social norms and obligations with a notion of proper economic conduct of all ranks within the local community. Scott (1976) extended the term and applied it to peasant households that live on the margin of subsistence, which makes them weavers of reciprocal webs of social obligation and dependence and supporters of rational economic principles based on judgements of social justice. The conceptual history of moral economy shows a great diversity of meaning; however, the term retains its analytical relevance: for instance, Götz (2015) considers civil society as employing economic options in moral ways (altruistic meaning of economic transactions) and encompassing the conversion of monetary and other resources into "moral capital". Thus, the term "moral economy" stands for (1) ethical norms to support each other in everyday social-economic practices, and (2) obligations for successful members of the community to take care of the weaker ones.

3 The English–Russian confusion of the terms "*narodnik* movement" and "populism" was already recognised in Soviet times (Khoros, 1980).

4 On 17 September, at a meeting with the heads of regional authorities, 70-year-old Savchenko unexpectedly announced his resignation from the position of governor. See, e.g., "The fall of Belgorod the Great with its endless day: The Governor Evgeny Savchenko resigned after 27 years of rule," *Kommersant*. 17.09.2020. Available from: www.kommersant.ru/doc/4494231 [viewed 25 June 2021].

5 For many years Grudinin refused to sell the land or the shares of his joint-stock company, "Lenin State Farm". He fended off all lawsuits, but today the company collects donations so that Grudinin can compensate his own enterprise for the "loss" of 1.66 billion rubles. According to the company's web-site, "Its workers and head do not have this money, which means that the shares of the enterprise would go to raiders. With the help of unfair courts, dishonest deputies and leaders, the raiders try to seize the lands and assets of the company. They have already succeeded in this many times, only near our settlement, there are two ruined agricultural enterprises overgrown with weeds and built up with multi-storey buildings. We cannot let this happen to our home! Help us save our company and prevent the construction of another multi-storey complex on its place!" Available from: https://sovhozlenina.ru/help-us [viewed 13 October 2020].

6 The Belgorod Regional Duma unanimously endowed Savchenko with the powers of its representative in the Federation Council. Savchenko headed the list of the United Russia in the elections to the Regional Duma on the Single Election Day in September. He was elected a member of the Legislative Assembly thus, receiving the right to apply for a senatorial mandate. See https://rg.ru/2020/09/22/reg-cfo/evgenij-savchenko-stal-senatorom-ot-belgorodskoj-oblasti.html [viewed 12 October 2020].

7   See, e.g.: "Non-small small deeds". Available from: https://selgazeta.ru/proektyi/
    selo/nemalyie-malyie-dela.html [viewed 14 October 2020]: "Today, the majority
    of municipalities and settlements are engaged in the implementation of large-
    scale, ambitious projects... However, there are also small deeds that improve the
    standards of living and the appearance of the territory, and contribute to the local
    civil activity and development of a comfortable living environment... Such small
    deeds can be exemplified by successful solutions by the local self-government of
    urgent problems of the settlement, which do not require large expenditures from
    the local budget, but are of high importance for the local population."
8   Available from www.ugory.ru [viewed 19 July 2020].

References

Amelina, M. (2000). Why Russian peasants remain in collective farms: A household
perspective on agricultural restructuring. *Post-Soviet Geography and Economics*,
41(7), 483–511.

Artemiev, M. (2010). Where the "red belt" disappeared [online]. [Viewed 19 July
2020]. Available from: www.forbes.ru/forbes/issue/2010-01-0/38461-kuda-ischez-
%C2%ABkrasnyi-poyas%C2%BB.

Badiou, A. (2016). Twenty-four notes on the uses of the word "people". In Badiou, A.,
Bourdieu, P., Butler, J., Didi-Huberman, G., Khiari, S., and Rancière, J., eds. *What
is a people?* New York: Columbia University Press, pp. 21–31.

Berelowitch, A. (2011). Putin's paths to populism. In Pugacheva, M. G. and Zharkov, V.P.,
eds. *Paths of Russia: Historization of social experience*. Moscow: New Literary
Review, pp. 216–223 (In Russ.).

Berelowitch, A. (2017). Putin is a populist compared to Western leaders. *Bulletin of
Public Opinion*, 1–2, 82–90 (In Russ.).

Bernstein, H. (2004). Changing before our very eyes: Agrarian questions and the politics of land in capitalism today. *Journal of Agrarian Change*, 4(1–2), 190–225.

Bozhenko, R. (2013). Melnichenko V.: "There is no such stupid thing that the authorities would not do". *Arguments and Facts*. 16.04.2013 (In Russ.).

Brubaker, R. 2020. Populism and nationalism. *Nations and Nationalism*, 2(1),
44–66.

Canovan, M. (1981). *Populism*. New York: Harcourt Brace Javonovich.

Chayanov, A. V. (1989a). On the theory of the non-capitalist economy. In Chayanov, A. V.,
ed. *Peasant economy. Selected works*. Moscow: Economica. pp. 114–143 (In Russ.).

Chayanov, A. V. (1989b). The journey of my brother Alexei to the land of peasant
Utopia. In Chayanov, A. V., ed. *Venetian mirror*. Moscow: Sovremennik. pp. 161–
208 (In Russ.).

Chayanov, A. V. (2006). Possible future of agriculture. *Economic heritage of
A. V. Chayanov*. Moscow: Tonchu. pp. 111–143 (In Russ.).

Cultural mosaic: Small territories – great opportunities, 2019. Collection of cases.
Moscow: Prospect (In Russ.).

Davydov, M. A. (2018). On the issues of the Russian modernization. *Russian History*,
3, 33–44 (In Russ.).

Deiwiks, C. (2009). Populism [online]. [Viewed 5 September 2020]. Available from:
https://ethz.ch/content/dam/ethz/special-interest/gess/cis/cis-dam/CIS_DAM_2015/
WorkingPapers/Living_Reviews_Democracy/Deiwiks.PDF

Edelman, M. (2005). Bringing the moral economy back in... to the study of 21st century transnational peasant movements. *American Anthropologist*, 107(3), 331–345.

Edelman, M. (2020). From "populist moment" to authoritarian era: Challenges, dangers, possibilities. *Journal of Peasant Studies* [online]. [Viewed 2 September 2020]. https://doi.org/10.1080/03066150.2020.1802250.

Edelman, M., Oya, C., and Borras, S. M. (2013). Global land grabs: Historical processes, theoretical and methodological implications and current trajectories. *Third World Quarterly*, 34(9), 1517–1531.

Fadeeva, O. P. (2019). Siberian village: From formal self-government to forced self-organization. *ECO*, 4, 71–94 (In Russ.).

Fedyunin, S. Yu., and Pain, E. A. (2019). Populism and elitism in contemporary Russia: Analysis of the relationship. *Political Studies*, 1, 33–38 (In Russ.).

Frühauf, M., Guggenberger, G., Meinel, T., Theesfeld, I., and Lentz, S. eds. (2020). *Kulunda: Climate smart agriculture. South Siberian agro-steppe as pioneering region for sustainable land use.* Switzerland: Springer.

Furman, F. P. (2009). The discourse of populism and the paradigm of the *narodnik* movement. *Humanities and Social-Economic Sciences*, 1, 113–115 (In Russ.).

Gordeeva, I. A. (2003). Forgotten people. *History of the Russian communitarian movement.* Moscow: AIRO-XX (In Russ.)

Götz, N. (2015). "Moral economy": Its conceptual history and analytical prospects. *Journal of Global Ethics*, 11(2), 147–162.

Granberg, L., and Sätre. A. M. (2016). *The other Russia: Local experience and societal change.* London: Routledge.

Gudkov, L. (2017). Peculiarities of the Russian populism. *Bulletin of Public Opinion*, 1–2, 91–105 (In Russ.).

Hall, S. (1985). Authoritarian populism: A reply to Jessop et al. *New Left Review*, 151, 115–124.

Harrison, M. (1975). Chayanov and the economics of the Russian peasantry. *Journal of Peasant Studies*, 2(4), 389–417.

Ionescu, G., and Gellner, E. eds. (1969). *Populism – its meanings and national characteristics.* London: Weidenfeld and Nicolson.

Ivanou, A. (2019). Authoritarian populism in rural Belarus: Distinction, commonalities, and projected finale. *Journal of Peasant Studies*, 46(3), 586–605.

Ivanov, P. (2019). Hands off the "Lenin State Farm" enterprise and its head Pavel Grudinin [online]. [Viewed 10 October 2020]. Available from: https://msk.kprf.ru/2019/06/11/115515 (In Russ.).

Karpov, A. E. (2017). Forum "Sustainable Development of Settlements" in the *Dobraya Zemlya* ["Kind Land" in the Vladimir Region]. *Russian Peasant Studies*, 2(4), 191–194 (In Russ.).

Khoros, V. G. (1980). *Ideological movements of the populist type in developing countries.* Moscow: Nauka (In Russ.).

Korolenko, V. (2018/1965). *The story of my contemporary*, 2 vols. Moscow: Vremya (In Russ.).

Kuznetsova, T. E. (2003). *Social thought on the natural potential of Russia: 12th century – Soviet period of the 20th century.* Moscow: Nauka (In Russ.).

Laclau, E. (2005). *On populist reason.* London/New York: Verso.

Latukhina, K. (2019). Harvest with a guarantee Vladimir Putin urged to take into account the interests of agrarians in each national project. *Russian Newspaper*. 26.12.2019 (In Russ.).

Lindner, P., and Nikulin, A. (2004). Everything around here belongs to the *kolkhoz*, everything around here is mine: Collectivism and egalitarianism: A red thread through Russian history? *Europa Regional*, 12(1), 32–41.

Magun, A. (2014). Protest movement of 2011–2012 in Russia: New populism of the middle class. *Stasis*, 2(1), 192–226 (In Russ.).

Mamonova, M. (2019). Understanding the silent majority in authoritarian populism: What can we learn from popular support for Putin in rural Russia? *Journal of Peasant Studies*, 46(3), 561–585.

McMichael, P. (2013). *Food regimes and agrarian questions.* Halifax/Winnipeg: Fernwood Publishing.

Moffitt, B. (2016). *The global rise of populism: Performance, political style, and representation.* Stanford: Stanford University Press.

Mudde, C. (2004). The populist zeitgeist. *Government and Opposition*, 39(4), 541–563.

Nikulin, A., Trotsuk, I., and Wegren, S. (2017). The importance of strong regional leadership in Russia: The Belgorod miracle in agriculture. *Eurasian Geography and Economics*, 58(3), 316–339.

Nikulin, A., Trotsuk, I., and Wegren, S. (2018). Ideology and philosophy of the successful regional development in contemporary Russia: The Belgorod case. *Russian Peasant Studies*, 3(1), 99–116.

Nikulin, A. M. (2020). *Chayanov school: Utopia and rural development.* Moscow: Delo.

Panizza, F. ed. (2005). *Populism and the mirror of democracy.* London: Verso.

Petrov, N., Makarenko, B., Denisenko, M. B., et al. (2013). *Russia 2025: Scenarios for the future.* London: Palgrave Macmillan.

Pivovarov, Yu. L. (2001). Urbanization in Russia in the 20th century: Presentations and reality. *Social Sciences and the Present*, 6, 101–113 (In Russ.).

Pokrovsky, N. E., and Nefedova, T. G. (2014). *Potential of the near north: Economy, ecology, rural settlements.* Moscow: Logos (In Russ.).

Pozanenko, A. A. (2019). Spatial isolation and sustainability of local communities: Development of existing approaches. In: Pugacheva, M. G., ed. *Paths of Russia. Boundaries of politics.* Moscow: NLO. pp. 139–153 (In Russ.).

Rodoman, B. B. (2017). Zoning as a way for possessing the space. *Regional Studies*, 3, 4–12 (In Russ.).

Roman-Alcalá, A. (2020). Agrarian anarchism and authoritarian populism: Towards a more (state-)critical "critical agrarian studies". *Journal of Peasant Studies* [online]. [Viewed 20 May 2020]. Available from: https://doi.org/10.1080/03066150.2020.1755840.

Savchenko, E. (2020). We must be in time! Speech at the World Russian people's council. *Our Contemporary*, 1, 91–98 (In Russ.).

Scoones, I., Edelman, M., Borras Jr., S. M., Hall, R., Wolford, W., and White, B. (2018). Emancipatory rural politics: Confronting authoritarian populism. *Journal of Peasant Studies*, 45(1), 1–20.

Scott, J. C. (1976). *The moral economy of the peasant: Rebellion and subsistence in southeast Asia.* New Haven: Yale University Press.

Shanin, T., ed. (1983). *Late Marx and the Russian road: Marx and "the peripheries of capitalism": A case.* New York: Monthly Review Press.

Shatskikh, V. (2016). Rural detective: Little-known life details of the famous farmer and orator Melnichenko [online]. [Viewed 10 October 2020]. Available from: www.solidarnost.org/promo/derevenskiy-detektiv (In Russ.).

Solovieva, E. V., Strunina, N. V., and Moiseeva, G. B. (2014). Russian populism is ineradicable. *21st Century: Results of the Past and Issues of the Present*, 4, 328–330 (In Russ.).

Taggart, P. (2000). *Populism*. Buckingham: Open University Press.

Thompson, E. P. (1971). The moral economy of the English crowd in the eighteenth century. *Past & Present*, 50, 76–136.

Tikhonova, V. V. (2011). *Political "foreleaders" and populism in contemporary Russia*. PhD thesis, Moscow State Regional University (In Russ.).

Titov, S. (2020). "Sometimes evil wins": Ex-presidential candidate Pavel Grudinin fights for his business [online]. [Viewed 19 July 2020]. Available from: www.forbes. ru/biznes/394995-zlo-mestami-pobezhdaet-eks-kandidat-v-prezidenty-pavel-gru-dinin-o-borbe-za-svoy-biznes (In Russ.).

Trotsuk, I. V., Nikulin, A. M., and Wegren, S. K. (2018). Interpretations and dimensions of food security in contemporary Russia: Discursive and real contradictions. *Universe of Russia: Sociology. Ethnology*, 27(1), 34–64 (In Russ.).

Uzun, V. Ya., and Shagaida, N. I. (2015). *Agrarian reform in post-Soviet Russia: Mechanisms and results*. Moscow: Delo (In Russ.).

Wegren, S. (2006). *The moral economy reconsidered: Russia's search for agrarian capitalism*. New York: Palgrave.

Wegren, S. K., and Konitzer, A. (2006). The 2003 Russian Duma election and the decline in rural support for the Communist party. *Electoral Studies*, 25(4), 577–595.

Wegren, S., Nikulin, A., and Trotsuk, I. (2018). *Food policy and food security: Putting food on the Russian table*. Lanham: Lexington Books.

Wegren, S. K., Nikulin, A. M., and Trotsuk, I. V. (2019). Russian agriculture during Putin's fourth term: A SWOT analysis. *Post-Communist Economies*, 31(4), 419–450.

Yudin, G., and Matveev, I. (2017). A politician without the people. Is it correct to refer to Putin as a populist? [online]. [Viewed 15 September 2020]. Available from: https://republic.ru/posts/82802 (In Russ.).

Zubarevich, N. V. (2011). Four Russias [online]. [Viewed 19 July 2020]. Available from: www.vedomosti.ru/opinion/articles/2011/12/30/chetyre_rossii (In Russ.).

Zverev, V. V. (1997). Evolution of populism: "Theory of small deeds". *Social History*, 4, 86–94 (In Russ.).

Zverev, V. V. (2016). Echoes of the "theory of small deeds" in the early 20th century. *RUDN Journal of Russian History*, 15(1), 29–38 (In Russ.).

7 The feeling of being robbed

Bjørn Egil Flø

Introduction

Norway is teeming with rural dissatisfaction these days. We have seen this clearly during recent years, and even more so in this last year. We have seen rural activists dressed in traditional folk dress rowing boats to campaign for the continued operation of delivery rooms; we have seen local communities rise in protest against closing regional schools and colleges and against the reforms of the municipality structure, the police and the postal service. At the same time, we have seen heavy artillery used in campaigns against toll booths in West Norway, and otherwise good-natured islanders from Frøya in Mid-Norway have smeared faeces on the footboards of excavators clearing the land for windmills belonging to multinational power companies. In brief, we have seen what ever more people are calling rural resistance. But where does this dissatisfaction come from, and what is at the roots of this rural resistance?

If we look behind the specific issues, behind all the reforms, behind the centralisation of the police and the downscaling of the postal service and educational facilities and what people see as cut-backs in basic infrastructure and welfare services where they live, we will see that people have started losing faith. If we look behind what comes across as resistance to the green shift, behind the demonstrations against toll roads and the anti-windmill campaigns, we will see that the grounds for this resistance is mistrust, and the grounds for this mistrust is betrayal.

In this chapter I want to communicate to readers some of the things that people in Norwegian rural areas have told me. I want to share my thoughts on what we today know by the relatively imprecise term "rural resistance". I want to explore what it originates in, and I will argue that it is a reaction against a development whereby people in rural areas lose out whilst people in urban areas are winners. I will argue that the resistance is directed at a neo-liberal modernisation advocated by an urban political, financial and cultural elite, reinforced by an ecological modernisation heavily founded on economic rationalist thinking, championed by the same class. While great parts of the ongoing socio-political research and "the pundits" – or the political commentators and commentary journalists – have presented the conflict in the rural–urban divide as socially constructed representations, I argue that it is

DOI: 10.4324/9781003091714-8

first and foremost about materiality, even though the aforementioned representations also play a part. It is about the very real loss of jobs and the disadvantages that local communities in rural areas are subjected to, without getting anything in return. And to the extent that representations play a part, it is about urbanity representing this loss. That is why people in rural areas reject the knowledge that is being produced, as well as the political statements put forward by the representatives of "the city centres".

It has been smouldering for a long time

Rural dissatisfaction has been smouldering for a while. And those who know rural Norway have probably seen it coming for a long time. People who have taken the time to talk to people in the local communities, up on the mountain farms and along the valleys, in the innermost parts of the fjords and farthest out on the islands, have seen that people there have started giving up hope.

For my part, it took time before I saw it. Only in the early 2000s, as I was wandering about in a small fisheries community in the Lofoten islands, did it start to occur to me. I had an appointment for an interview with an entrepreneur in the field of tourism and the meeting was postponed, so I took the opportunity to have a look around this once-upon-a-time so lively fisheries community. The last time I had been there was in 1984. At the time I had a leave of absence from military service (conscripted in Norway) and went there in the hope of seeing what I thought were the last remnants of Norwegian whaling. But it struck me that what I was looking at back then, in the 1980s, were the last remnants of a local community.

"Yes, things have changed a lot since back then." The man of advancing years came to a halt and looked right at me. It was as if he doubted me. "What did you say you were doing here?" he asked. "I'm here in connection with a research project on tourism." I could tell by his face that his doubts were confirmed. "Yes, that seems to be what everything is about now – tourists and leisure," he said, with a look brimming with mistrust of an industry that already at the time had been hyped as the Great Deliverance.

For me, the talk with the retired whaler was a wake-up call. Since I started doing rural research in 1999, this was the first profound conversation about what I at the time thought were the perceived drawbacks of tourism, but which I later understood was a critique of the new rural policies.

He told me about a dying local community that only 20 years earlier had been brimming with life. When I was there in the 1980s, there were more than 400 people living there. The small fishing boats lay three abreast at the quay and you could hear people swearing and laughing at the fish landing centre and the dispatch area. It sounded like it is supposed to sound when people are at work, and it smelled of fish, petrol and sea. But in 2004, little of it was left. The population had shrunk to 180 people, and many feared that what still remained of livelihoods, places to meet, service functions and welfare services would disappear in the near future.

"Tourism alone can't make up for this," said the retired fisherman. "For that we need politics, we need rural policies and fisheries policies that ensure that the value creation from the fisheries benefits the local communities."

After talking to several hundreds of people in local communities in the context of various research projects over the years, I noticed that they have increasingly been telling such stories. They have been talking about local communities that have gradually been emptied of jobs and of venues for meeting up. They talk about ever more impoverished communities, about acquisitions and the closing down of local businesses, and they see it as an active downscaling of rural development measures to their own detriment and in favour of market forces. I have listened to stories about people out in the rural areas feeling disempowered and alienated. They perceive that the agents of change are outside themselves, outside their community, outside their region and even outside the country. "They are in the globalization of finance and culture", as Reidar Almås (2002, p. 408) put it. People who work with production have less say than ever before, especially in the primary industries. "An ever smaller part of the value creation comes about in the primary industries and ever more of it comes about with the efforts of capital, foreign raw materials, scientifically based knowledge and external distribution" (Almås, 2002, p. 408). So they're left there, the farmers, the fishermen and the villagers, stuck in a manufacturing chain of factor inputs and processing industry segments, feeling small and disempowered. Because in the driver's seat are forces outside and beyond both the rural areas and the primary industries.

The cultural turn and a downscaled rural policy

Rural dissatisfaction has, however, rarely been the subject matter of the journalists and the commenting experts in the national media. There, content relating to rural policies has been relegated to the feature departments, who produce reports such as the weekend supplement to one of Norway's leading newspapers, with the title *Bygdestyret*, a pun on "rural conflict", a term meaning a kind of rural version of "the tall poppy syndrome" and "rural rule" in Norwegian (A-Magasinet, 2017), or the report in the weekend supplement to the largest Norwegian financial newspaper (DN-Magasin, 2014) with the even cleverer title *Lande-klage*: a pun on "rural complaint", "plague" and "hit" in Norwegian. It tends to be the exotic side of rurality that is highlighted in such articles. A reporter and a photographer typically go to a rural area to report on a foreign culture, writing articles that are beautifully illustrated with photos of café interiors, such as Klætt kafé in Klætt or the Konditoriet in Rjukan, interiors every bit as much part of the retro style fad in interior decorating, with their functionalist Nordic design teak and Respatex furniture from the 1960s, just like apartments in the capital featured in interior design magazines. Yes, there have been good articles, written by skillful journalists, but they have rarely prompted political debate. They have seldom been mentioned in the editorials on page two, because the rural areas are no longer considered to be an issue of political interest.

But maybe one cannot blame the editors alone; maybe this just reflects the general debate on social issues? My own observations and the above-mentioned reflections by Almås (2002) in many ways broke with the then ruling perspective among academics in general and rural sociologists in particular. Because at some time in the early 1990s, the formerly so materially oriented rural sociology abandoned its traditional heavily socio-economic perspective for more social constructivist and social representation-theoretical perspectives (Bell, 2007; Flø, 2009, 2013). And precisely this shift – or the cultural turn, as the British human geographer Paul Cloke (1997) calls it – had consequences for the development of rural areas as well as for the rural policies (Perkins, 2006; Flø, 2010, 2013, 2015). Gradually, Norwegian officialdom let go of the sturdy measures that had actually worked for so long. "The development of infrastructure in the rural areas was put out to tender with price and short-term profits as the main criterion," says a former local politician from Nordland county. "And the term 'redistribution' practically disappeared from the vocabulary of our most prominent politicians." The cultural turn can be linked to the shift in rural policies in the 1990s, a shift that according to Håvard Teigen (2020) ended up as a "merged-up and downscaled" rural development policy. Inspired by post-modern theories on the new consumption (Best, 1989; Perkins, 2006), more and more people started hoping that status-pursuing middle-class consumers, employed in liberal professions and with more money and leisure time than ever before, would take a liking to rurality (Flø, 2013). Scientists, consultants and politicians opined that rural areas should start making themselves attractive to the growing urban middle class. Farmers should make use of what they saw as niches in various markets and start "crafting good stories" (content marketing) and "wrapping" their products in local and environmentally friendly symbols:

> Rurality was to be sold. Rural traditions, activities and culture was be turned into commodities and sold in this new market, and farmers and villagers were to be taught to stimulate the consumers' hunger for symbol products lending the consumers distinction.
>
> (Flø, 2015, p. 20)

With substantial support from the Norwegian Hospitality Association (*NHO Reiseliv*) and Innovation Norway, rural areas were to be turned into what Aasetre (2010) calls "recreational colonies" for a steadily growing middle class on the look-out for trendy identities to self-project on social media.

From social democratic order to neoliberal order

"It started in the 1980s," said the thin, sinewy older man. He heaved a box of fish from his small fishing boat onto the quay, before leaping ashore. "It started with those new, slippery politician types of ours, the ones with the shoulder pads and long hair at the back of their necks," he continued. He had voted for them himself, he said. Because he had believed in them, for a long time.

Maybe the fisherman has a point. Maybe you have to go back to the 1980s to find the source of the rural dissatisfaction. Maybe you have to go back to the political scene after Thatcher and Reagan, politics that Thatcher's highly trusted cabinet member Nigel Lawson (2011) described as "a mixture of free markets, financial discipline, firm control over public expenditure, tax reductions, nationalism, Victorian values, privatisation and a dash of populism" (Chapter 5, paragraph 9).

The historian Berge Furre (1992) also highlighted the political shift that took place in the 1980s. "The social democratic order was in the main dismantled in the eighties," writes Furre (p. 421). The social democratic order was typically characterised by mixed market economy compromises and a strong belief in infrastructure and regulations. The concern of the rural areas, with ancestry going back to the early days of industrialisation, was still very much alive for our politicians. "They carry their history about with them," a long-term mayor and later deputy MP told me in 2006. He was talking about the older politicians in his own party, the Norwegian labour party. "They remembered our early members of parliament, like Johan Castberg, and the substantial attention to rural policies that characterised our early institution building."

The former politician and judge Johan Castberg is a good example. Reading the minutes from the parliamentary debate on the concession laws in 1917, this attention to rural issues comes across very clearly:

> We shall to the greatest possible extent take the opportunity to strengthen these more remote and barren areas by letting them have a substantial part of those values that are created within their borders... You take the values out of the rural areas and channel them into the industrial centres and the cities, emptying the villages, emptying the local communities, especially the mountain communities, acquiring their values without remunerating them and this is absolutely wrong.
>
> (quoted in the White Paper Norwegian Official Report
> (*NOU*), 2012: 09, p 70)

The heritage from Mr Castberg and several other of our early MPs is what I have formerly called the heritage of a "morally rooted" attention to rural issues (Flø, 2017). But what transpired during the discussions on the concession acts was in itself also a heritage. It was the heritage of "the Norwegianist movement" (*norskdomsrørsla*), part of the nation-building in Norway during the 19th century, after 400 years as a Danish colony came to an end when Sweden sided with England during the Napoleonic wars and subsequently demanded Norway as a "war prize" to be handed over by the Danish king in 1814. The arrangement was called a union. The union with Sweden was dissolved, unilaterally, by Norway in 1905. The "Norwegianist movement" and the general nation-building period focused largely on rural musical, literary, artistic and folk dress traditions – and, of course, language. That heritage, with all the songs saluting mountains and fjords, the paintings depicting beautiful scenery in remote places, presented important symbols for the

Norwegianist movement. The cabinet-making inspired by folk crafts, the music inspired by folk music traditions and the poems and the literature from that period, a lot of it written in a Norwegian aiming to relieve itself of the Danish language formerly imposed on Norwegian writers, stayed with the politics of the interwar period and the crisis compromise between the Labour Party and the then "Farmer's party", a party that also in its present form and with a new name (the Centre Party) continues to be concerned about decentralisation issues and rural areas. This heritage informed rural policies during the entire duration of the social democratic order.

Another former mayor, from the North Gudbrandsdalen valley, was of the opinion that our "new politicians" were incapable of understanding the needs of the municipalities, along with those of rural industry and commerce – especially farming: "We could have tough fights before as well, but back then, in the 80s and the 90s, it was like banging your head against a wall." Arguing well for something didn't get you anywhere, because "they just didn't get the reasons argued", he said, explaining that also in earlier days MPs and cabinet members could "be stingy", but he never doubted that "they understood the realities out there in the municipalities". Implicit in this story is a sense that the mayor feels that "they" – meaning "the new politicians" – ignore the municipalities in rural Norway. The moral position that the rural districts had had in the political debate under the social democratic order had been weakened with "those new and slippery politician types", as the fisherman at the quay had put it. At any rate, this was what one of the local activists against oil production in the fisheries hub where the Lofoten and Vesterålen islands are was referring to when he in 2017 voiced the opinion that "we have a Minister of Petroleum that is f---ing useless", before elaborating "This is not going to become Stavanger!" By that he was referring to a doctrine that has a strong presence in the Norwegian writing of history – i.e. that of the political strategist and long-time mayor of Stavanger, Arne Rettedal (Conservative), who made sure that the depressed area of Stavanger acquired a special position in the development of the Norwegian petroleum industry (Kindingstad, 1998). "Oil won't bring any more jobs to North Norway, nobody thinks like that about developing industry anymore, the oil will only gobble up the jobs we already have", the activist continued. And just before Christmas 2019, the news that Equinor will not be landing oil and gas from the Castberg oil field locally was released. It was hardly a surprise that the mayor of Nordkapp (the North Cape) feels cheated.

From the late 1980s, it seems that the attention to rural areas was increasingly pushed into the background, with a greater focus directed towards business development policies – at least, that is what I hear from people in rural areas. And even if Berge Furre (1992) hesitated to give the new era the contours of a name like the era before it – i.e. the social democratic order – I will name this era "the neoliberal order". The neoliberal order is very much alive today and is, like Nigel Lawson (2011) says, marked by a naïve belief in a laissez-faire economy, privatisation, tax cuts, deregulating, free trade and cuts in public expenditure.

Deregulating and new regulating hand in hand

During the last 25–30 years, the attitude that publicly run enterprises are not sufficiently efficient became established in the general political debate. The attitudes of the 1950s, '60s and '70s have been turned upside down. The scepticism towards the governance-enthusiastic state was guided by economists, hotshot players in the business community and the influential newspaper and television editors who were blooming so profusely before the Internet became public property. We get little in return for our tax money, they opined, and so it was time to open the door to the private business sector.

The critical attitude to the publicly run operations was hardly rooted in tax resistance and public inefficiency, says former senior civil servant and CEO of Telenor Tormod Hermansen (2004). It is also about the publicly run operations having invaded areas that traditionally belonged to the private sphere and the civil society.

That was also how the fisherman on the quay justified that he for a long time had voted for these kinds of politicians. "Everything was bureaucracy, overruling of local decisions and the introduction of principles that were out of place here in the High North," he explained to me, in an attempt to justify voting for parties that he was no longer proud of having voted for. He had been stopped when he wanted to build a small house for equipment storage and a quay in order to fish directly from his home, instead of having to rent a house for equipment and tools a three-hour drive from where he lived. "The littoral zone act," he said simply, and as he said it, I knew that he was referring to part of the Norwegian Planning and Building Act (*plan- og bygningslova*) (2008, sections 1–8) and the prohibition against building in the 100 metre littoral zone – or 100 metres from the shore – which came in 1989. Similar reasoning also transpires in a study I did in the district I come from (Flø, 2013). Here, there was an old barn that the farmer was not allowed to tear down because the chairperson of the county counsel thought it fitted beautifully into a tourism landscape that the bureaucrats were vividly imagining. "You know, that's the sort of thing that makes us do so well in the opinion polls," my friend told me, referring to the right-wing Progress Party's (FrP) steadily increasing support in Sunnmøre county in the west of Norway in the early 2000s. In my notes, I have many similar stories that all testify to Hermansen (2004) having a point. "The man in the street", whose opinions right-wing populists so often claim to voice, felt that public institutions got too close and became too "hands-on" as regulators. They perceived the paradox of a period marked by deregulation in many fields also carrying in it new regulations in others, especially in the field of environmental issues, and gradually more and more so as regards the field of climate.

Sustainable development

After the environmental discourse of the late 1960s and well into the 1980s had been dominated by pessimistic prophets who had read up on classics such as *Limits of Growth* (Meadows, Meadows, Randers, & Behrens, 1972) and *Blueprint for Survival* (Goldsmith & Allen, 1973), the dystopian world view that the Western consumer society suffered from a deep cultural crisis was created, and this dystopian view was substituted by an idea of turning our social and environmental problems into the business model itself. Because that, apparently, was what it was going to take. If we were to have a hope of any "sustainable development", it had to be profitable. That was the great promise.

With the final report from the Brundtland Commission (World Commission on Environment and Development, 1987), the term "sustainable development" became universal. But something that was also turned into a universally accepted concept was that the changeover to a more sustainable society should not change the basic principles of the economy. To the contrary, in order to succeed in accomplishing the environmental objectives, we were to preserve the existing economic regime. This way, our understanding of sustainable development contributed to the continued economic growth essential to success in the work before us. At the same time, we excommunicated all dissidents and thought that they were naïve utopians.

Still, there were a lot of things to like about the Brundtland Commission and the term "sustainable development". Here, the local grassroots participation was emphasised, and with a diversity of decentralised, experimental approaches, sustainable development was turned into a democratic strategy for environmental protection, rather than one governed by experts. Many people, such as political scientist Maarten Hajer (1995), considered the term sustainability to be "one of the paradigm statements of ecological modernisation" (p 26) and a forerunner of the perspective that now dominates the discourse on the environment and the climate – i.e. ecological modernisation.

Ecological modernisation
Sustainable development and ecological modernisation have several common features. Both perspectives see the need for more cross-sectoral environmental politics. Both think that economic growth is not incompatible with solving the environmental problems. Both consider technology to be an important means of solving environmental problems.

Later on (i.e. from the last half of the 2000s and to the present), it seems as if technology is no longer just a means of solving environmental challenges: it has become the end itself. Here is the business concept, here are the grounds for favourable financing of research and development as well as for patenting and establishing industries and here is the competitive edge in the fight for the clients.

All the same, there is a difference between sustainable development and ecological modernisation. The ecological modernisation perspective is not overly concerned about distribution and fairness, asserts Oluf Langhelle (1998). And maybe that is exactly what we are looking at now? Maybe we are looking at the consequences of the fact that ecological modernisation has unfairly distributed impacts. When rebels wearing *Gilets Jaunes* (yellow vests) set fire to cars along the Champs-Élysées, it is a sign of a lack of social sustainability. When people damage toll booths in Bergen and pile faeces onto the footboards of excavators on Frøya, it is a sign that we have wandered far off course on the way to a socially sustainable green shift.

Because that is what they are protesting against. That is what the resistance to toll roads is really about. That is what the anti-windmill activists on Frøya and people along most of our coast are so furious about. They are not climate deniers or opposed to a green shift, they are just opposed to the way that shift is attempted brought about. They are opposed to what Robyn Eckersley (1995) calls "the discipline of environmental economics' key role in shaping the economy-environment integration debate" (p 12). They are simply opposed to most of what is meant by the term "environmental economics".

Environmental economics

Environmental economics are based on a neo-classical socio-economic theory. This is the academic foundation for that "environmental discourse" that John Dryzek (1997) called "economic rationalism" – i.e., the principle of using market mechanisms in order to accomplish public objectives. This constituted the grounds for privatising nature worthy of preservation. This became particularly popular in the USA in the 1980s, when business-like principles were introduced for the American national parks, turning them into a hospitality industry, whereby visitors pay to come and see the national park and the profit is used for the maintenance of it (Buckley, 2003). Likewise, this also constituted the grounds for the "pollutor pays" principle and the introduction of green taxation and environmental taxes. And it is this principle that Hajer (1995) thinks has created a technocratic version of ecological modernisation permeated with financial reasoning. But the principle stuck, and since the late 1980s this has been the most legitimate way to discuss environmental politics, internationally as well as nationally.

This was also what triggered the resistance. The resistance came about as a consequence of the measures resulting in unfair impacts. The logic of the environmental economics impacted some more heavily than others. And the resistance is not irrational, contrary to what the commenting journalists and the people in expensive shirts keep telling us. It is, rather, a direct and rational reaction based on basic ethics of fairness. Road tolls, fuel tax and higher parking charges impact people in outlying areas hardest. These expenses impact those who, from the 1950s and well into the 1980s, did exactly what they were encouraged to do. They impacted people who bought land in the developments that the municipalities had put up for sale in the small communities and the peripheral areas outside the densely populated areas and cities. They impacted those who for a long time had called themselves social democrats and had believed in the promises made by politicians. They impacted those people who for the longest time had had the most confidence in the politicians elected to be their representatives, the same politicians who strove to reduce public expenses by centralising the schools and the public services. Now they are stuck in their small communities and their monocultural housing developments – 20 to 40 km or 12 to 24 miles from a town or city centre – knowing that there is nothing there but a bed and a kitchen. Everything else has to be got in the town or city centre. The school is shut down, the shop is closed and what was once a living fish landing centre has been taken over by a self-made turbo-capitalist who swore that he would ensure the future operation of it, but ended up landing his fish elsewhere, thereby depriving the place of all economic activity.

New regional policies

In 1992, parliament passed a resolution to merge what translate as "The Regional Development Fund" (*Distriktenes utbyggingsfond*), "The Industrial Fund" (*Industrifondet*) and "The Small Businesses Fund" (*Småbedriftsfondet*) into the National Industries and District Development Fund (*Statens nærings- og distriktsutviklingsfond*, SND). The White Paper with the historical title translating as "City and Country, Hand in Hand" (a Labour Party slogan from 1933 which rhymes in Norwegian) (St. meld. Nr. 33 (1992–1993)) was termed by Håvard Teigen (1999) as a White Paper "of great historical interest because the regional politicies unique to Norway are discontinued with this White Paper" (p. 203). At the end of the 1990s and on into the 2000s, many of the business and industry policies and the economic development policies were delimited by the EEA agreement, and the opportunities to use classical business development measures such as cheap credit, cheap power, interest-free loans and other kinds of crisis support measures were more or less removed. What we often call globalisation came ever more to the fore for large parts of the economic activities, especially in the rural areas. An ever closer economic, political and cultural integration into Europe and the rest of the world curtailed the space for traditional Norwegian economic development measures, and gradually the more neoliberal currents also caught up with the rural areas.

During the three or four years before the financial crisis in 2008, the most remote municipalities lost more than 6,000 inhabitants. In spite of the so-called "red–green" coalition (the Labour Party, the Socialist Popular Party and the rurally oriented Centre Party) being self-proclaimed rural development enthusiasts, 210 of the then 430 municipalities in the country shrank considerably during this period. And while rural areas had been home to more than half of the Norwegian population in the late 1980s, to day seven out of ten Norwegians live in, or near, the biggest cities.[1]

Some regions noticed the changes very clearly into the 2000s. While some regions – like the Sunnmøre region, where I come from – did well as a consequence of business owners insisting on developing their business locally, other regions were practically closed down. And the manufacturing and processing industries we had then have now moved to parts of the world where costs and taxes are lower. But the local business owners have their challenges, because hardly anybody wants to work in the manufacturing industry anymore; nowadays, young people dream of becoming "something in the media". And if somebody was interested in becoming a skilled worker in the manufacturing industry, they would probably be diagnosed as spineless underachievers by school counsellors and others who queue up to advise young people to move away from the community they come from (Brox, 2016). A former CEO at the Kleven shipyard in Ulsteinvik told me, in 2004, that "there is nothing we would rather do than recruit local young people, but it's impossible to get hold of them [...]. Nowadays they're all going to university." The solution for local business owners was to import workers from EEA countries. The influx of capable skilled workers from Eastern Europe was followed by a deterioration of working conditions and wages. In January 2020, the Kleven shipyard became Croatian-owned, with the conglomerate DIV Group in the driving seat. Some of the employees felt relieved at having avoided bankruptcy, while others counted the days to a feared closure. "They could have given us a green assignment," said a former shipyard colleague of mine. "The Prime Minister could have asked us to build these offshore windmills, I know we could have done it." But she didn't. Premier Solberg and the other politicians we have entrusted with our vote no longer ask for things like that; they pin their faith on the markets. And in the meantime our skilled workers pack up and move, and get replaced by temporary workers conveyed by staffing agencies.

Our production industries become distant from the local communities that they were once a part of. Young people get professional training or education that requires them to move away from rural Norway as there are no jobs for them there. The regional university colleges seem to be a thing of the past, and the vocational schools attach increasing importance to theoretical subjects. This growing emphasis on theoretical subjects leads Knut Kjelstadli to note that the great number of dropouts from the vocational schools is due to the fact that they have let the students down through creating a "a school for reading and listening" rather than "a school for seeing and doing" (Skolen, 2006). This all contributes to changing our industries as well as our local communities.

Big structural changes, especially in the fisheries (Isaksen & Bendiksen, 2002), have had extensive local spillover effects that made a lot of people in rural areas every bit as concerned as the man I met in Lofoten. Everybody worries about what they see as the destitution of their local communities. They look at their homes and their communities and see them die a little more every day. The broad rural policies, meaning rural policies that include "weighty" political fields such as defence, manufacturing industry and transport – a type of vision that has been the lynchpin for Norwegian rural policies for decades – are now unravelling. It is now the *narrow* rural policies, emphasising individual measures in order to entice individual players to start businesses in rural areas rather than in cities, that country people should pin their faith on.[2] Nowadays, the rural areas are to be built with "project finance".

From the early 2000s onwards, we could read reports and in-depth interviews with young, newly hired project managers who were working on making the local communities more attractive for families with small children. The political counterculture that was typical of the anti-urbanites moving to the rural areas in the 1970s has now been replaced with a smooth compliance culture, the projection of identity and staged rural idyll. The inherent message from the many move-back-home projects in various parts of the country was "If you are resourceful and energetic, move to our area". Like mediocre travel agencies, rural Norway tried to sell itself by the means of advertising materials financed by Innovation Norway, striking the chords of light entertainment connected to outdoor recreation and relaxation.

The background for this was a political vision wherein natural and cultural values were emphasised as ever more important effort factors for new businesses. It was a policy based on a desire to facilitate the establishment of creative businesses and thereby create local communities that came across as more attractive, especially to young people. The key words were adventure, identity and symbols of distinction. Or, as the then Minister of Municipalities and Regional Affairs, Åslaug Haga, put it when she presented the government's action plan for culture and business in June 2007: "By focusing on [...] culture, adventure and leisure, this will contribute to rendering a place attractive and stimulating a sense of identity. Culture and adventure means a lot to people, regardless of whereabouts they live" (Norwegian Ministry of Trade and Industry, 2007). Haga had the support of both then Minister of Trade and Industry Dag Terje Andersen and then Minister of Culture Trond Giske, who had both participated in working out the action plan.

This was the new regional policy, and it must be understood as being part of a bigger whole. It is permeated with the thinking of the academics and the suppliers of knowledge to the players involved with regional policies. In keeping with the spirit of the times, attempts were made to turn local communities into attractive tourism destinations, with waiters dressed in national costume offering visitors local delicacies. The clearest shift in the neoliberal order was an individualisation of the responsibility for your own success. The business community was relieved of the burden of regional policies, and in rural areas the project creators were given the responsibility for regional development.

The fight for survival is hardening

In the course of approximately the last hundred years, we have moved on from morally rooted regional policies, via regional policies in party politics marked by mixed economy responses to political and market-related challenges, before stepping into a neoliberal order where the state has worked hard at washing its hands of the responsibility for rural development and left an ever-greater part of it to the local communities, local businesses and the locals themselves. Downscaled regional policies, professionalised politicians and an active trimming of the nation-state in favour of international agreements and global players was followed by the betrayal that fed the growing dissatisfaction in rural districts where people saw the consequences right before their eyes. They saw their resources being harvested, they heard the sounds of lorries that were empty on arrival and left fully loaded, day and night, but they saw little of the value created by the exports.

It is in this context that we should understand the man I met in the Lofoten islands, and a whole lot of other people in rural areas I have talked to in recent years. Be they dairy farmers in Telemark or MPs representing a district, former mayors from the north of the Gudbrandsdalen valley, sheep farmers in the north of the Østerdalen valley, county politicians in Helgeland up north, hairdressers in fjord hamlets down south, blue-collar workers in Sogn og Fjordane county in the west or business owners in Sunnmøre – they are all watching their own local community, and what they are seeing is that the fight for survival is hardening. Many people have a feeling similar to what a smallholder in West Norway explained to me: "We have seen the robbery happening before our very eyes for a long time now. And here they're now going to produce power for the Brits without paying us."

They have felt the shifts in regional policies at their cost. Local fish landing centres, dairies, abattoirs and boatyards are disappearing, along with shops, schools and public health centres. And "the only thing we're offered is a White Paper on tourism," as one of my informants put it, referring to White Paper (*Stortingsmelding*) No.19 (St. meld. Nr. 19 (2016–2017)), which told them to turn culture, adventure and leisure into business. And they are perceiving this all as a betrayal.

All the same, they roll their sleeves up and take hold of what they can. They stake capital and health on building the country; they invest millions in required loose housing barns and buy cod quotas at exorbitant prices. They enter into partnerships with R&D players for developing innovative welding technology and new methods for the recycling of waste from the building and construction industry as well as from the fisheries industry. All this and more is what I see when I travel through the local communities, but on television the ever-debate-ready know-it-alls lecture about how the local communities have to adapt and facilitate the implementation of the promising bioeconomy and the green shift.

Rural people are not asking for advice about adapting, they are just asking us to see that adapting is exactly what they are doing. They are in the business of adapting every single day; every single day they are working steadily on creating value, jobs and welfare. They produce materiality: food and goods for export. And they produce a lot of collective goods: cultural landscapes, biodiversity, food safety and living local communities – goods us city people can enjoy on our way to the chalets worth millions that we have spread along the littoral zones and in the best summer mountain pastures. What they have seen of the bioeconomic industry so far differs little from the fossil-economic industry. It sprouts ever more distant multinational players who come to steal the wind as well as the kelp and the spruce forests without leaving anything in return but disadvantages for the people living there.

Us and the others

And what do we do? We don't want to see them. We – in David Goodhart's (2017) understanding of *anywheres* as the urban middle-class elite with higher education – dominate the public debate with our socio-liberal "citizen of the world" identity. We ignore their work and underestimate their contributions as well as their feelings, invalidating them and calling it "nostalgic sentimentalism". We accuse them of being backwater luddites holding the opinion that all change is loss.

Ourselves, we claim that we are devoid of nostalgia because we are egalitarian meritocrats who are of the opinion that we have deserved the powers we hold because of our own intelligence and reason. We therefore close down local hospitals and delivery rooms; we close down local police stations and call it "the community police reform"; we dismiss all local referendums that oppose municipal amalgamations and claim that people do not know what is in their own best interests. We demand that the farmers, who have shown us a higher growth in productivity than practically any other industry, must become more efficient. And should any of us come up with the clear-headed idea of moving a job or two in state enterprises to a "forgotten small town in some rural area", the whining from the bigger cities drown all other news for weeks.

We are still seeing only the early beginnings of a rural resistance. It will be a long time before a possible rebellion materialises as anything more than legal protest marches, support for certain political parties and civil remarks in newspapers and social media. But precisely because of that, it might be time to slow down. Maybe it is time that we – as representatives of the urban middle class with higher education who dominate the political scene, the public administration and the public debate – start looking around us, start listening. Because what we need now is for us as a society anew to be able to discover the mutual benefits between city and countryside and realising that we need to look up from the short-term economic models, see the bigger picture and acknowledge that the rural areas have their own intrinsic value at the same level as the cities. Only then might we possibly manage to relieve the feeling of robbery in the rural areas.

Notes

1 Statistics Norway SSB.
2 The divide between the narrow and the broad rural policies is the same divide that Håvard Teigen (1999, 2020) defines as explicit and implicit rural policies. Explicit rural policies consist of regionally differentiated political measures, whereas implicit rural policies are the sum total of allocations that distribute and re-distribute funds between the regions. Allocations of funds to the municipalities over the government budget make up the most important part of these measures, but transport, defence, industry, agriculture and other political fields are also significant elements in the implicit rural policies.

108   Bjørn Egil Flø

References

Aasetre, J. 2010. En rekreativ koloni: Bygdeutvikling på middelklassens premisser. *Plan* (5), 58–62.

Almås, R. 2002. *Frå bondesamfunn til bioindustri: 1920–2000* (Vol. IV). Oslo: Det Norske Samlaget.

A-Magasinet. 2017. Bygdestyret. *Hilde Lundgaard*, 17 February.

Bell, M. M. 2007. The two-ness of rural life and the ends of rural scholarship. *Journal of Rural Studies*, 23 (4), 402–415.

Best, S. 1989. The commodification of reality and the reality of commodification: Jean Baudrillard and post-modernism. *Current Perspectives in Social Theory*, *19*, 23–51.

Brox, O. 2016. *På vei mot et post-industrielt klassesamfunn?* Oslo: Pax Forlag.

Buckley, R. 2003. Pay to play in parks: An Australian policy perspective on visitor fees in public protected areas. *Journal of Sustainable Tourism*, *11*(1), 56–73.

Cloke, P. 1997. Country backwater to virtual village? Rural studies and "the cultural turn". *Journal of Rural Studies*, *13* (4), 367–375.

Dn-Magasin. 2014. Lande-klage. *Ole Øyvind Sand Holte*. 6 September 2014.

Dryzek, J. 1997. *The politics of the earth. Environmental discourses*. New York: Oxford University Press.

Eckersley, R. 1995. *Markets, the state and the environment: Towards integration*. South Melbourne: Macmillan Education Australia.

Flø, B. E. 2009. Vondtet i norsk bygdeutvikling. *Syn og Segn* (3), 76–80.

Flø, B. E. 2010. Bygda – forståing og implikasjonar. *Plan* (5), 20–25.

Flø, B. E. 2013. Me og dei andre. Om lindukar, Framstegspartiet og bygda som sosial konstruksjon. *Sosiologisk Tidsskrift*, *21* (2), 152–168.

Flø, B. E. 2015. *Bygda som vare – om bygda, elgen og folkeskikken*. Trondheim: NTNU.

Flø, B. E. 2017. Det grøne skiftet og grendene – sentrum og periferi i den nye økonomien. *Plan*, 2, 26–29.

Furre, B. 1992. *Norsk historie 1905–1990. Vårt hundreår* (Vol. 4). Oslo: Det Norske Samlaget.

Goldsmith, E. and Allen, R. 1973. *A blueprint for survival*. New York: Penguin Books Ltd.

Goodhart, D. 2017. *The road to somewhere: The populist revolt and the future of politics*. London: Hurst.

Hajer, M. A. 1995. *The politics of environmental discourse – Ecological mordernization and the policy process*. Oxford: Clarendon Press.

Hermansen, T. 2004. Den nyliberalistiske staten. *Nytt Norsk Tidsskrift*, *21* (3–4), 306–319.

Isaksen, J. R. and Bendiksen, B. I. 2002. *Fiskeindustrien i Troms og Finnmark – Strukturendringer og verdiskaping* (No. 18/2002): NOFIMA.

Kindingstad, T. 1998. *Maktens byggherre. Historien om politikeren, ingeniøren og småbrukeren Arne Rettedal*. Stavanger: Wigerstrand.

Langhelle, O. 1998. *Økologisk modernisering og bærekraftig utvikling – to sider av samme sak?* Report (No. 4/98): ProSus.

Lawson, N. 2011. *Memories of a Tory radical* (Kindle version). London: Biteback Publishing. Retrieved from amazon.com

Meadows, D. H., Meadows, D. L., Randers, J. and Behrens, W. W. (1972). *The limits to growth: A report for the Club of Rome's project on the predicament of mankind*. New York: Universe Books.

Norwegian Ministry of Trade and Industry (Nærings- OG Handelsdepartementet). 2007. Regjeringen lanserer handlingsplan for kultur og næring, *Pressrelease*. Oslo.

Norwegian Official Report (NOU) 2012. *Energiutredningen – verdiskaping, forsyningssikkerhet og miljø*. Oslo: Olje- og energidepartementet.

Perkins, H. C. 2006. Commodification: Re-resourcing rural areas. In P. Cloke, T. Marsden and P. H. Mooney (eds.), *Handbook of rural studies*, 243–257. London: SAGE Publications.

Planning and Building Act (Plan- og Bygningsloven). 2008. *Lov om planlegging og byggesaksbehandling*. 27 June.

Skolen, I. 2006. Intervju med Knut Kjeldstadli. *Anonymous*, 9, 4.

St. meld. Nr. 19. 2016–2017. *Opplev Norge – unik og eventyrlig*. Nærings- og fiskeridepartementet.

St. meld. Nr. 33. 1992–1993. *By og land hand i hand. Om regional utvikling*. Kommunal- og regionaldepartementet.

Teigen, H. 1999. *Regional økonomi og politikk*. Oslo: Universitetsforlaget.

Teigen, H. 2020. Distriktspolitikkens historie: Frå nasjonsbygging til distriktsopprør. In R. Almås and E. M. Fuglestad (eds.) *Distriktsopprøret. Periferien på nytt i sentrum*. Oslo: Dreyers forlag.

World Commission On Environment and Development. (1987). *Our common future*. Oxford: Oxford University Press.

Part II
Policies of rural authenticity

8 #Proudofthefarmer

Authenticity, populism and rural masculinity in the 2019 Dutch farmers' protests

Anke Bosma and Esther Peeren

Introduction

On 1 October 2019, hundreds of tractors clogged the motorways leading to The Hague, where a big farmers' demonstration took place. The tractors caused a major disruption and the largest traffic jams in Dutch history (NOS, 2019a). In the following months, more demonstrations and blockades by farmers followed. The protest wave came after a tense summer, which began when, on 13 May, animal rights activists occupied a pig farm in Boxtel, demanding to film the conditions under which the pigs were being kept and to evacuate any sick or injured animals. The political response to the occupation was overwhelmingly critical, with Minister of Agriculture, Nature and Food Quality, Carola Schouten stating: "This is one of the businesses providing our daily food. They must be able to do this without constantly having to look over their shoulder and being intimidated" (Huijben, 2019).[1] The events of 13 May led to the founding of two new farmers' organisations: the Farmers Defence Force (FDF), which aims to fight back against "the excesses of environmental extremists" while comparing itself, in its willingness to take radical action, to Greenpeace; and Agractie, dedicated to representing farmers' interests and connecting farmers and citizens, but also prepared to engage in direct action.

The immediate impetus for the 1 October protest, which involved both the FDF and Agractie, was a statement by MP Tjeerd de Groot of the social-liberal D66 party that the number of pigs and chickens in the Netherlands should be cut in half in order to comply with EU regulations on nitrogen deposits. Despite the fact that De Groot excluded cows, goats, sheep, horses, rabbits, ducks and turkeys from his plan, it was presented in the Dutch media as aiming for a reduction by 50% of all livestock, and was taken by many farmers as posing a direct threat to their livelihoods (Winterman, 2019). De Groot's political opponents rushed to side with the farmers and denounced the plan. MP Gert Jan Segers of the Christian progressive party CU summarised it in a television interview as aiming to "take farmers down a peg" (Nieuwsuur, 2019).

Coming so soon after the Boxtel pig farm occupation and in a context of widespread anger among Dutch farmers at negative media coverage of

DOI: 10.4324/9781003091714-10

agricultural issues and changing government regulations, De Groot's plan may have been the catalyst for the protest, but it was not its only or even most important motivator.[2] Significantly, one of the organisers of the 1 October protest, sheep farmer Bart Kemp, stated in an interview with regional newspaper *De Gelderlander* that the main aim was to prompt a long overdue reappreciation of farmers (Van Essen, 2019a). To achieve this, the protesters sought to present themselves as hardworking guardians of the food supply (through the slogan "no farmers, no food"), unjustly beleaguered by politicians and misunderstood by a general public concentrated in urban areas.

In this chapter, we analyse the role played by a particular notion of authenticity in the discursive framing of the Dutch farmers' protests of late 2019 and early 2020 by the protesters and various politicians, as well as in the remarkably sympathetic initial public response.[3] It is our contention that the authenticity claimed by and ascribed to the protesting farmers drew legitimacy from the intimate association of authenticity with the rural identified and critiqued by the German philosopher Theodor Adorno in *The Jargon of Authenticity* (1973). The fact that the rural in general, and farmers in particular, have long been taken as exemplifying authenticity and thus as inherently authentic has made the "authentic farmer" a pleonasm. This ingrained idea, we will show, was not only what drove the early support for the farmers' protests by a large majority of the Dutch public, but what also facilitated an alignment between the protests and a populist-nationalist politics.

The idea that farmers are inherently authentic and thus people to be proud of (a sentiment expressed by the popular protest-related hashtag #proudofthefarmer, from which we take our title) relies, as we will make clear by way of Adorno, on an idyllic image of farming that has little to do with its realities and is therefore open to contention. At the beginning of the protests, the farmers successfully evoked this idyllic image by presenting themselves as running vulnerable family farms, intimately connected to their land and animals, and working 24/7 to feed the nation not for financial reward but out of pure dedication. However, public support declined when news reports began to appear that challenged this framing.[4] These reports eroded the "perceived authenticity" (Chhabra, Healy, & Sills, 2003) of the protests, revealing them to be "strategically staged" (MacCannell, 1973) by political actors making dubious statements.

First, it was reported that the number of Dutch farms had already declined from 410,000 in 1950 to 54,000 in 2018 because of the long-standing policy of scale enlargement and that, in the same period, livestock numbers had grown rapidly, with pigs increasing from 2,407,000 in 1956 to 48,971,000 in 2018, laying hens from 33,024,000 to 47,302,000 and, most staggeringly, broilers from 2,407,000 to 48,971,000 (CBS, 2019). This undermined the farmers' claims that there would be an unprecedented reduction in the number of farms if livestock numbers were halved and that Dutch farming was still characterised by small-scale family businesses. Second, the fact that 75% of Dutch agricultural produce is exported (Leijten, 2019) undercut the idea that all existing farms are needed to feed the nation. Third, the repeated claim that farmers work extremely hard for very little money was challenged by Statistics Netherlands (CBS) researcher Peter Hein van Mulligen, who tweeted that 18% of Dutch farmers are millionaires (Phvanmulligen, 2019). Since, according to Lionel Trilling, "money [...] is the principle of the inauthentic in human existence" (Trilling, 2009, p. 124), pointing out that not all farmers are barely able to sustain themselves and their families made the protesters seem less authentic and consequently less deserving of sympathy. It also cast doubt on whether the protesters, who had stressed farmers' financial struggles, had "accurately represent[ed] th[eir] values and commitments" (Jones, 2016, p. 492), sowing the seeds for accusations of hypocrisy. Fourth, press coverage revealing that the farmers' protests had received substantial financial support from agricultural multinationals such as slaughterhouse company Vion and fodder companies De Heus and ForFarmers (Verhoeven, 2019) challenged the communicated image of the protests as a grassroots effort and of farmers as paragons of self-sufficiency.

Our main focus in this chapter is on how the farmers' protests were so successfully presented and eagerly accepted as authentic and therefore justified, especially at their start but also later on (almost half of the Dutch public continued to support the farmers). As noted, we argue that this was facilitated by the continuing traction of the idea of the inherently authentic farmer. Now that this idea is increasingly being used, in the Netherlands and in other countries, to buttress nationalist-populist politics, it is especially urgent to understand and connect its historical and current manifestations. Accordingly, we will first use Adorno's work to specify the long-standing link between farming, authenticity and nationalism, before showing how this link was revived, by certain Dutch politicians and some protesting farmers, to align the protests with a populist, anti-immigration politics. Subsequently, we draw on the work of Sara Ahmed and Michael Kimmel to explore how this alignment also involved the association of the protesting farmers with a particular form of rural masculinity that marked their anger and violence as innate and therefore authentic and justified. Finally, our conclusion details how, during the 2020 COVID-19 lockdown, a video made by Agractie once more appealed to the idea of the authentic farmer to rebuild public support for the farmers, suggesting that people's investment in this idea and its mobilising power persist.

Farming, authenticity and nationalism

One notion repeatedly evoked by the Dutch farmers to frame their protests was that farming is not something you do, but something you *are*. Thus, in an article on the website of the national television broadcaster NOS, 27-year-old dairy farmer Jelle Treurniet, who hopes to take over the farm started by his great-grandfather, was quoted as saying: "Being a farmer is part of you. If you disagree with farming policy, you cannot suddenly say that you do not want to do it anymore" (NOS, 2019b). Similar statements that "a farming business is not just a job, it's in your genes" and that "you either are a farmer or you are not one" were made by other farmers interviewed about the protests (Boerderij, 2019; Nieuwsuur, 2019), as well as by Wageningen University researcher Melle Nikkels in the national newspaper *Trouw*: "For agriculturists, 'being a farmer' is their whole identity. Their work is also their way of life" (Van Velzen, 2019b). The idea of farming as an essential, internal identity – and therefore not something that can easily be given up – evokes what philosopher Somogy Varga calls the "ideal of authenticity," defined as the sense that "one should lead a life that is expressive of what the person takes herself to be" (Varga, 2011, p. 1). If farmers take themselves to *be* farmers, this ideal prescribes that they should do everything possible to continue to live as they believe farmers should live.

In "The Paradox of Authenticity," Varga distinguishes three western models of authenticity that he argues followed each other historically: a quantitative, a qualitative and a performative one. The quantitative one, exemplified by the 19th-century works of Samuel Smiles and John Stuart Mill, relates the idea of self-realisation that is central to authenticity to the internalisation and "realization of something (at least thought to be) generally human" (Varga, 2011, p. 4). This model of authenticity is about being in tune with social norms regarding what it means to be what Smiles, in his 1859 *Self-Help*, calls "a man [*sic*] of character" (Varga, 2011, p. 3). Qualitative authenticity, in contrast, is about achieving harmony with "a unique, introspectively accessible core self" (Varga, 2011, p. 4). It is a 20th-century individualised model of authenticity that relies on the idea of a person realising herself as what she already *is* deep inside: an essential core that is considered invariable. Finally, Varga argues that the 1990s saw the emergence of a performative authenticity that conceives of authenticity as "developed in (strategic) interaction with others" and manifesting as "the *energy of difference*" (Varga, 2011, p. 6, emphasis in text). It "no longer refers to an inner teleology, but instead to a process of self-creation or difference-creation" that is without an identifiable origin and subject to change (Varga, 2011, p. 6). Its main aim is to differentiate the individual or collective self from others.

At first glance, the ideal of authenticity evoked by the farmers and identified by Nikkels seems to accord with the qualitative model of authenticity. However, we want to suggest that it also fits the quantitative model in referring to an essential inner core that is not individual but collective. Rather than being a unique identity, "being a farmer," as Nikkels emphasises, is a shared identity predicated on internalised social norms that equate being a farmer with being a "man of character". For Treurniet, moreover, being a farmer is an identity that is externally performed as different from and more authentic than other identities, as becomes clear when, in the same interview, he notes: "Those men with those ties in The Hague don't know anything about our business. They constantly tell us what to do, and it is always something different" (NOS, 2019b). This illustrates how his self-creation as an authentic farmer who could not be anything else is simultaneously an act of performative difference-creation through which he distinguishes himself from government politicians and bureaucrats. The latter are marked as inauthentic because instead of being what they do, like farmers, their jobs are considered external to their being, like the ties they wear. In addition, they are associated with a changeability ("it is always something different") that contrasts negatively with the fixed inner core of the authentic farmer.

Thus, the discourse of farming as an identity evoked in the framing of the protests combines elements of the quantitative, qualitative and performative models. This allows farmers in general to be presented as conforming to the ideal of authenticity, as living a life that expresses not just who they take themselves to be, but also who a large part of the Dutch general public, participating in upholding the social norm of the farmer as a hardworking "real" person, takes them to be. Because the notion of farming as in and of itself an authentic activity is widely shared in the Netherlands, the ideal of authenticity in this case implies that nothing should be done by anyone – including the government – to prevent farmers from living their farming lives.

The deep roots of the connection between farming and authenticity are revealed in Adorno's *The Jargon of Authenticity*. There, he argues that the German existentialist notion of authenticity, championed most prominently by Martin Heidegger in his 1927 *Being and Time*, relies on "the silent identification of the archaic with the genuine" (Adorno, 1973, p. 51). In taking farmers and shepherds as exemplifying the archaic and therefore the genuine, Heidegger is seen to ascribe a "false eternity" to "agrarian conditions" and to reify "transitory social forms which are incompatible with the contemporary state of the forces of production" (Adorno, 1973, pp. 56, 47). According to Adorno, the jargon or cult of authenticity that emerged from, among others, Heidegger's account shows an "ignorance of everything we have learned about rural people" in identifying rural dwellers as rooted and therefore authentic, and urbanites as dispersed and therefore inauthentic (Adorno, 1973, p. 55). This overlooks not only the hardship and exploitation that capitalism imposes on small farmers, putting their livelihoods in "perpetual crisis", but also how the lack of prospects drove the younger sons of German farmers to commit "the worst atrocities in the concentration camps" of World War II (Adorno, 1973, pp. 55, 27). In Adorno's view, the Heideggerian jargon of authenticity, in its emphasis on a rootedness that is identified as specifically "Germanic-Swabian" and associated with purity, endogamy, inwardness, self-sufficiency and a distaste for the unknown, underpinned the fascism of Nazi Germany (Adorno, 1973, p. 54).

Heidegger was able to invest farming with an inherent authenticity on the basis of an ahistorical, idealised perception of the rural that was not unique to him but can be traced back to the ancient genre of the rural idyll. As the Russian literary theorist Mikhail Bakhtin explains, the idyll, across its different historical forms, conjures a fantasy of a limited, autonomous space in which everything is familiar and nothing ever really changes as time moves in a cyclical pattern, with each generation engaging in the same activities and maintaining the same worldview (Bakhtin, 1983, pp. 224–225). This fantasy easily accommodates the type of nationalist space of rootedness, homogeneity and timelessness that Heidegger saw the rural as embodying.

Fernando Molina and Antonio Miguez Macho supplement Adorno's and Bakhtin's work by highlighting the crucial role the rural idyll, as an "ethno-romanticist rural imaginary" that exceeded the literary realm, played in the emergence of nationalism across Europe in the 19th century (Molina & Macho, 2016, p. 688). They argue that the "rural archetype" forged by the nation-builders of this era (which included both politicians and artists) "transmitted the essential values of the nation: permanence, biological purity, common ancestry. These values fed into the most basic discursive structure of nationalism, turning the peasantry into a living icon of the nation as an ancestral community" (Molina & Macho, 2016, p. 688). In the contemporary context of the Dutch farmers' protests, we will show next, the long-standing, intimate link between an idyllic notion of farming, the idea of authenticity and an exclusive nationalism was reactivated both by populist, anti-immigrant politicians who aligned themselves with the protests and by the protesters themselves.

Protesting farmers as populist heroes

At the 1 October protest, Tjeerd de Groot's speech, in which he tried to justify his proposal to curtail livestock numbers, was drowned out by the protesters' jeers. The same happened when the leader of the progressive green party Groen Links spoke about the importance for the environment of limiting nitrogen emissions (NOS, 2019a). A much more positive reception, however, was given to the speeches made by two populist-nationalist MPs: Geert Wilders, leader of the PVV (Party for Freedom), and Theo Hiddema, MP for the FvD (Forum for Democracy), led by Thierry Baudet. Wilders climbed onto a tractor and framed the farmers as emblems of nationalism: "You are the heroes of the Netherlands. Where would the Netherlands be without our farmers? You provide our food … What you deserve is not hassle, but respect" (Geertwilderspvv, 2019). Hiddema, a lawyer who grew up on a farm, was interviewed on the main podium and professed that his "farmer's heart" was "warmed" by the protests (Forum voor Democratie, 2019a). His message that Dutch farmers had been fooled by mainstream political parties for too long and therefore should continue to protest was greeted with loud applause and cheers. At a follow-up protest in The Hague, on 16 October, Baudet joined Hiddema to reinforce the FvD's message of support. On this occasion, Hiddema explicitly invoked the idea of farmers as inherently authentic and thus superior to other groups by referring to them as "people with a real job and real worries" (Forum voor Democratie, 2019d). On 9 and 11 October, moreover, Baudet spoke in support of the protesting farmers in Parliament.

Tellingly, Baudet's 11 October speech is presented by the FvD on YouTube under the title "The City Versus the Country". The speech is grounded in the nationalist "dichotomous dynamic" that Molina and Macho summarise as "country/tradition/moral purity versus city/modernity/corruption" (Molina & Macho, 2016, p. 689). It takes as its starting point the conservative British philosopher Roger Scruton's argument that industrialisation has uprooted people from an organic engagement with the earth, the land and animals. Crucially, Baudet does not link this lost organic engagement to a nature untouched by human hands (which he considers a romanticised delusion on the part of animal rights and climate change activists), but to husbandry (Forum voor Democratie, 2019c). In the 9 October debate, Baudet heralds the *boerenstand* or peasantry as "constituting our connection with the earth, with the seasons and with the animals" and as shaping "the Netherlands that we know, the Netherlands that we associate with our youth and our history," which, for him, is the Netherlands as it was painted by Frans Hals and Vincent van Gogh (Forum voor Democratie, 2019b). Here, the dehistoricised "rural archetype" Molina and Macho describe is enlisted to invest present-day Dutch farmers, many of whom run large-scale, heavily automated and digitised businesses, with the supposed naturalness and authenticity of pre-industrial, feudal peasants.

Moreover, Baudet's insistent use of "we" and "our" in the speeches is designed to exclude immigrants, who cannot claim roots in what he designates as "our splendid (at) home" (Forum voor Democratie, 2019b). That the FvD's glorification of farmers as upholders of the Dutch nation is designed to reinforce its demonisation of immigrants is made explicit in an interview, where Baudet calls the peasantry the "last bastion against state bureaucracy" and argues that the disempowerment of farmers is intrinsically linked to the pressure put on "the land" by "mass immigration" (Boerenbusiness, 2019).

It was not just that Baudet and Wilders strategically linked their populist-nationalist anti-immigration politics to the protests; such a politics, based on the "ritualized demonization of an 'other' seen as unravelling the threads that weave together the idealized unified 'traditional' national culture and the core ethnic stock" (Berlet & Sunshine, 2019, pp. 480–481), was also mobilised by some of the protesters. Thus, FDF board member Sieta van Keimpema, instead of denying the nitrogen crisis, shifted the blame for it onto immigrants:

> The nitrogen problem grew because of immigration. Every human breathes out nitrogen; if there are more humans, that will therefore cause an increase. Immigrants also bring with them their own consumption patterns. That [the foreign products immigrants want to consume] must come from somewhere. Population growth therefore also produces nitrogen.
>
> (quoted in Goudsmit & Straver, 2020)

In addition, at a December 2019 protest in Hilversum, a Confederate flag – often presented as a general symbol of rebellion, but widely linked to white supremacy – was flown from a pick-up truck, while a homophobic text was painted on the back of a caravan implying that those on the left side of the political spectrum are "not real men" (Mascini, 2019). In the next section, we expand on the role the notion of the farmer as a "real (heterosexual) man" played in the framing of the protests.

The righteous anger of the farmer as a "real" man

The idea that "modern urban civilization had a feminizing effect", which has been prevalent in western thought since the late 19th century (Kimmel, 2017, p. 47), identifies the rural as a masculine realm. According to Jo Little, moreover, "physical aggression, and strength" are central to the type of masculinity associated with the rural, which is "performed through control of the landscape" (Little, 2002, pp. 480–481). Farming, as the major form such control has historically taken and as traditionally (but not necessarily presently) involving heavy and dirty physical labour, is considered the epitome of rural masculinity. The heavy machinery involved in modern farming further contributes to its masculine image.

The idea of farmers as "real men" was central to a protest-related campaign launched in November 2019 by farmers' organisation Team Agro NL. It sought to get the song *De boer dat is de keerl* [The farmer that is the man] by the Dutch band Normaal voted to a high spot in the Top2000, an annual list of 'best songs ever' broadcast on public radio (Aalbers, 2019). Normaal's songs are known for expressing rural pride, and the band has been on a "quest to redeem the farmer and the rural Netherlands" since 1975 (Zwiers, 2015, p. 86). *De boer dat is de keerl* ended up in the number 9 spot, underlining the sympathy many Dutch people felt for the protesting farmers (Aalbers, 2019). The song's title equates farmers with masculinity, suggesting that farmers, more authentically than any other group, embody manliness. Furthermore, this is a superior kind of manliness, as the farmer is not just *a* man, but *the* man, with *keerl* specifically indicating a manly man.

Tellingly disregarded by the song, and also partially in the protests, is the possibility that a woman could be a farmer. In Dutch, instead of both sexes being included in the word "farmer," a gendered derivation is used: the male farmer is a *boer* and the female farmer a *boerin*. The way in which all the main protest-related hashtags and slogans used *boer* consigned women farmers to invisibility. Moreover, even when they did become visible in the protests, they were presented as women with masculine traits or a "manly" devotion to fighting for the cause. An interview with the aforementioned Van Keimpema in the national newspaper *De Volkskrant*, for example, pictured her holding a rifle (Schoorl, 2019), while another newspaper published an article headlined "Sieta van Keimpema a Militant? That Depends on Who You Ask" (Ruitenbeek, 2019). It is no coincidence that Van Keimpema's militancy was questioned – not just in terms of whether she *was* militant, but also in the sense of whether, as a woman, she *should be* militant – while the combativeness of the male protesters was taken as part of their authentic farmer selves and therefore considered legitimate.

In *Angry White Men*, Michael Kimmel describes how emotional vulnerability in men – feelings of anxiety, sadness, grief and worry – can be transformed into political rage. He argues that this is most successfully done by populists, as "populism is more an emotion than it is an ideology. And that emotion is anger" (Kimmel, 2017, p. 8). Showing emotions may not conventionally be what "real men" do, but this does not go for all emotions, with masculine anger in particular being considered "socially inherited and justified" and even "innate" (Helal, 2005, pp. 80, 88). In the terms of Varga's qualitative model of authenticity, male anger – and the violence flowing from it – can thus be taken as a sign of men being in harmony with their core self (Varga, 2011, p. 4).

The figure of the "angry white man" is common in present-day populist politics throughout the western world. Kimmel explains its emergence by referring to "aggrieved entitlement": a feeling of having lost or being on the verge of losing control and power in areas where white men expect to have – and feel entitled to – dominance. In the American context, on which Kimmel focuses, it indicates "the sense that 'we', the rightful heirs of America's bounty, have had what is 'rightfully ours' taken away from us by 'them,' faceless, feckless government bureaucrats, and given to 'them,' undeserving minorities, immigrants, women, gays and their ilk" (Kimmel, 2017, p. 32). As our discussion of the presence of anti-immigration rhetoric, racist symbols and homophobic slurs in and around the Dutch farmers' protests has shown, this way of thinking also played a role there. The protesting farmers framed themselves – and were framed by politicians such as Baudet – as the rightful heirs of the bounty of the Netherlands, threatened by environmental and animal rights activists and government bureaucrats – in the guise of "those men with those ties in The Hague" that farmer Treurniet complained about – and in danger of losing out to (non-white) immigrants. Aggrieved entitlement, Kimmel shows, allows angry white men to claim victimhood and dominance at the same time. This is also true of the protesting farmers, who framed themselves and were framed by populist politicians as beleaguered victims, but also as maintaining the capacity to dominate culturally, as the guardians of "the Netherlands that we know," and materially, by blocking highways with their imposing tractors and causing damage to protest sites and government buildings (Winterman, 2019).

That this framing was not recognised as strategic, but taken as expressing who the Dutch farmers really, authentically *are* is clear from the initial acceptance of the anger of the protesters by a large proportion of the Dutch public as legitimate. Even when the protests turned violent, there was relatively little condemnation from the public or politicians (Gargard, 2019). It seems that, because the farmers were perceived as "real" Dutchmen, they were seen to have the right to express their anger in materially destructive ways and to claim dominance.

In "The politics of good feeling," Sara Ahmed argues that "good things" are considered sources of happiness both because "happy" emotions get stuck to them and because there is a shared orientation towards these things as being good or as bringing happiness. She presents the white, heterosexual, patriarchal nuclear family as an example of a "good thing". We contend that the initial public reaction to the farmers' protests illustrates how, in the Netherlands, farmers have also been constructed as "good things" or "good people", enabling them to get away with violent acts of protest and expressions of anger that would not be tolerated from other groups. As Ahmed emphasises, the expression of anger does not necessarily inspire a negative response, but will do so when anger is expressed in a certain way by people regarded as not participating in a society's shared orientation towards things generally considered as being good. Thus, feminists who do not see the patriarchal nuclear family as bringing happiness are not seen as legitimately angry about women's oppression but as "kill joys" (Ahmed, 2008, p. 5). Furthermore, by expressing anger that is not considered legitimate, such feminists can become the point of tension in a room so that "the exposure of violence becomes [perceived as] the origin of violence" (Ahmed, 2008, p. 6).

This mechanism was arguably activated in the response to actions by Extinction Rebellion in late 2019, which included a traffic blockade in Amsterdam by protesters lying down in the street. Not only did the blockade lead to 90 protesters being put in jail for several hours (Verbeek, 2019), but the public and political response was a lot more negative than that to the farmers' protests. Politicians from centre-right parties and from the FvD and PVV denounced Extinction Rebellion in particularly strong terms. By calling out the violence being done to the planet, livestock and plant life, the environmental activists, much like the occupiers of the Boxtel pig farm discussed at the beginning of this chapter, became points of "bad feeling" conceived as "disturb[ing] the promise of happiness" (Ahmed, 2008, p. 6). Indeed, one seasoned Extinction Rebellion protester explained that when people criticise her, she is often told to "take into account other people's feelings" (De Ruiter, 2019). In contrast, the protesting bodies of (white) male farmers on tractors, which also caused disruptions and broke laws, were not (or not as widely) seen as generating "bad feeling." Notably, at the 1 October protest, only seven farmers were arrested (NU.nl, 2019).

A link can be made to authenticity here in the sense that the altruism represented by animal rights and environmental activists – who protest primarily on behalf of others (animals, the planet, future generations) – may be less readily perceived as authentic than self-interest, given that "authenticity's focus on being true to oneself seems to privilege one's desires over others" (Jones, 2016, p. 491). In addition, unlike the protesting farmers, who benefit from a social context in which "being a farmer" is equated to being an authentic "man of character", these activists are not perceived as authentic merely by virtue of their activism. In fact, environmental activists are often represented as hypocrites – and thus as inauthentic – in news reports and blogs that point out how they use products that are bad for the environment (Piotrowski, 2017, p. 845). This was precisely what CDA politician Diederik Boomsma did when he noted that because police helicopters had to be deployed, the Extinction Rebellion protest had caused extra CO_2 emissions (Verbeek, 2019). In addition, whereas the protesting farmers were constantly characterised as "hard workers" or, in Hiddema's words, "people with a real job", the Extinction Rebellion activists were, according to a tweet by Wilders, "leftist professional deadbeats, jobless climate hippies and ecological layabouts" and thus inauthentic protesters (Verbeek, 2019). By adding that such inauthentic protesters "naturally feel at home in Amsterdam", Wilders replicated Heidegger's mapping of the authentic/inauthentic divide onto the rural/urban. In the end, it was the Dutch public's shared perception of farmers as inherently authentic "men of character" or righteously angry "real (heterosexual) men" which ensured that, for a while, their protests received broad support.

Conclusion

After the initial wave of protests in late 2019, the farmers' organisations were invited to discuss the nitrogen emissions plans with the government, but when talks stalled, another protest followed on 19 February 2020. The police made clear that, this time, they would not tolerate tractors on the highways, but some farmers disobeyed, leading to a number of arrests (RTL Nieuws, 2020). By mid-March, the Netherlands had gone into lockdown to stop the spread of COVID-19. During the lockdown, Agractie released a video message addressed to the Dutch public (Agractie, 2020). It starts with the sound of claxons and an image of a long procession of tractors on a road over which the hashtag #OurFarmersOurFuture appears. A voice-over states: "You know us from the farmers' protests of the past months…" Then, the screen goes black and the klaxons stop. The voice-over continues, over images of food being harvested and a few cows eating, while sentimental piano music plays:[5] "But not now. Now our farmers are working hard to provide honest and safe Dutch food for everyone and to keep our businesses alive in this crisis. For the future of our Netherlands." The video ends by superimposing the Agractie logo on a soft-focus image of a boy feeding some cows. On the logo, it says "Our sector is at stake".

The video's narrative, and especially the stark transition from the klaxons and the imposing tractors to the piano music and reassuring images of harvesting and feeding cows, suggests a certain atonement for the disruption caused by the farmers' protests. To an extent, the video also breaks with the idyllic image of farming as small-scale and relying on the human labour of the farmer, as it portrays large harvesting and sorting machines, farm workers and an immense field being ploughed. However, the video does not show any intensive animal farming, the type of farming considered to require most scrutiny from animal rights and environmental perspectives; nor does it show any chicken or pigs, the proposed reduction of whose numbers was the main catalyst for the protests. In addition, the video reaffirms the framing of the protests through the link between authenticity, farming and nationalism. Its emphasis on the farmers' hard work and on the honesty and healthiness of the food they produce to sustain the Dutch people re-associates the farmers with an idyllic imagination of the rural as a realm of authenticity central to preserving the nation – a realm of authenticity that is implicitly contrasted with the inauthenticity of COVID-19 as a threat from abroad. Within this renewed frame of authenticity, the video reinforces the farmers' claim that everyone in the agricultural sector is having trouble keeping their head above water in "this crisis", which may be read as encompassing both COVID-19 and the proposed livestock reduction that precipitated the farmers' protests. While it is not yet clear whether the video succeeded in reinvigorating public support for the protests, which gained new momentum in June 2020, its message suggests a continuing confidence in people's susceptibility to a discursive framework that ties farming to authenticity and nationalism.

It was this susceptibility, evinced in the initially overwhelmingly positive public response to the protests, that sparked our investigation and led us to identify a particular notion of authenticity with long-standing links to the rural and nationalism as underpinning both the almost reflexive public support for the protests and their rapid embedding in a nationalist-populist "politics of the rural" (Woods, 2006) grounded in the reinvention of the rural–urban divide as one between "the people" and "the elite" (Mamonova & Franquesa, 2019). What our examination of the case study of the Dutch farmers' protests has highlighted is that fantasies like those of the inherently authentic farmer and the rural idyll, because of their long-standing affective overdetermination, which causes people to remain attached to them despite their lack of grounding in actuality (Peeren, 2018), remain highly effective political mobilisers. As long as the engrained idealised perception of farming and the rural that underlies these fantasies is not definitively dispelled, important rural policy discussions are easily derailed and populist narratives will continue to seem commonsensical to many.

Acknowledgement

This publication emerged from the project "Imagining the Rural in a Globalizing World" (RURALIMAGINATIONS, 2018–2023), which received funding from the European Research Council (ERC) under the European Union's Horizon 2020 research and innovation programme (grant agreement No 772436).

Notes

1 All translations from Dutch are the authors'.
2 In 2018, the national newspaper *Trouw*, in collaboration with Wageningen University, conducted the largest ever opinion poll of Dutch farmers. The 2018 results showed more than 80% of respondents agreeing that "farmers work hard, but barely receive any recognition" and that "in the media farmers are always blamed" (Trouw, 2019).
3 A poll presented on the RTL talk show *Jinek* put public support for the protests on 1 October at 89% (Jinek, 2020).
4 The poll presented on *Jinek* showed support for the protests falling to 60% on 16 October and 49% on 19 February (Jinek, 2020). Another RTL poll showed support dwindling from 70% in October to 55% in December, with 54% considering the disruptions caused by the protests unjustified in December, compared to only 34% in October (RTL Nieuws, 2019).

5 The piano plays a slow, conventional harmony in which any tension caused by dissonant chords is immediately resolved to create a reassuring and comforting listening experience. Research has shown that this type of music, when used in commercials, results in consumers perceiving a brand as "softer, more reserved, devoted and gentle" (Zander, 2006, p. 474).

References

Aalbers, R., 2019. Achterhoek zet Normaal unaniem op 1 in Top 2000. *De Gelderlander*, 25 December [online]. Available at: www.gelderlander.nl/bronckhorst/achterhoek-zet-normaal-unaniem-op-1-in-top-2000~a8448d29f/ (Accessed: 18 August 2020).

Adorno, T. W., 1973. *The jargon of authenticity*. Evanston: Northwestern University Press.

Agractie, 2020 [online video]. Available at: https://agractie.nl/ (Accessed: 14 July 2020).

Ahmed, S., 2008. The politics of good feeling. *ACRAWSA e-journal*, 4(1), pp. 1–18.

Bakhtin, M., 1983. Forms of time and of the chronotope in the novel. In Bakhtin, M., *The dialogic imagination: Four essays*. Austin: University of Texas Press, pp. 84–258.

Berlet, C. and Sunshine, S., 2019. Rural rage: The roots of right-wing populism in the United States. *The Journal of Peasant Studies*, 46(3), pp. 480–513.

Boerderij, 2019. "Boer zijn is iets heel groots, het is mijn plicht daarvoor op te komen". *Boerderij*, 11 November [online]. Available at: www.boerderij.nl/Boerenleven/Artikelen/2019/11/Boer-zijn-is-iets-heel-groots-het-is-mijn-plicht-daarvoor-op-te-komen-495392E (Accessed: 14 July 2020).

Boerenbusiness, 2019. Baudet: "Boeren moeten hun gang kunnen gaan". *Boerenbusiness*, 14 October [online]. Available at: www.boerenbusiness.nl/video/10884335/baudet-boeren-moeten-hun-gang-kunnen-gaan (Accessed: 14 July 2020).

CBS, 2019. Landbouw; vanaf 1851. *Statline*, 19 November [online]. Available at: https://opendata.cbs.nl/statline/#/CBS/nl/dataset/71904ned/table?ts=1571744094220 (Accessed: 14 July 2020).

Chhabra, D., Healy, R. and Sills, E., 2003. Staged authenticity and heritage tourism. *Annals of Tourism Research*, 30(3), pp. 702–719.

De Ruiter, M., 2019. Deze actiegroep pakt het klimaatprotest radicaler aan. *De Volkskrant*, 4 October [online]. Available at: www.volkskrant.nl/nieuws-achtergrond/deze-actiegroep-pakt-het-klimaatprotest-radicaler-aan-iedereen-zal-ons-haten-maar-horen-zullen-ze-ons~beaa3d79/ (Accessed: 14 July 2020).

Forum voor Democratie, 2019a. Hiddema: laat je niet langer bij de neus nemen! [online video]. 1 October. Available at: www.youtube.com/watch?v=q7M5b8UTmHs (Accessed: 14 July 2020).

Forum voor Democratie, 2019b. Baudet vs de Tjeerd de Groot (D66) [online video]. 9 October. Available at: www.youtube.com/watch?v=Wy4ClAeCyS0 (Accessed: 14 July 2020).

Forum voor Democratie, 2019c. *De stad versus het land* [online video]. 11 October. Available at: www.youtube.com/watch?v=_kFzpwf1Zb4 (Accessed: 14 July 2020).

Forum voor Democratie, 2019d. *Niet te geloven! Baudet en Hiddema bij het boerenprotest!* [online video]. 16 October. Available at: www.youtube.com/watch?v=vlI3rLco4Gw (Accessed: 14 July 2020).

Gargard, C., 2019. Wat de boze boer en de klimaatactivist met elkaar gemeen hebben. *De Correspondent*, December. Available at: https://decorrespondent.nl/10802/wat-de-boze-boer-en-de-klimaatactivist-met-elkaar-gemeen-hebben/3042583936348-16cf627c (Accessed: 14 July 2020).

Geertwilderspvv, 2019. *PVV STEUNT ONZE BOEREN! #boerenprotest #boerenprotest #boeren* [Twitter]. 1 October. Available at: https://twitter.com/geertwilderspvv/status/1178960515957219328 (Accessed: 14 July 2020).

Goudsmit, R. and Straver, F., 2020. Hoe Farmers Defence Force zijn "strijders" mobiliseert. *Trouw*, 19 February [online]. Available at: www.trouw.nl/duurzaamheid-natuur/hoe-farmers-defence-force-zijn-strijders-mobiliseert~bbc4182b/ (Accessed: 14 July 2020).

Helal, K. M., 2005. Anger, anxiety, abstraction: Virginia Woolf's "Submerged Truth". *South Central Review*, 22(2), pp. 78–94.

Huijben, M., 2019. Einde aan urenlange bezetting boerderij in Boxtel. *Brabants Dagblad*, 13 May [online]. Available at: www.bd.nl/meierij/einde-aan-urenlange-bezetting-boerderij-in-boxtel-tientallen-demonstranten-opgepakt~a945b4a8/ (Accessed: 14 July 2020).

Jinek, 2020. Steun voor boerenprotesten drastisch afgenomen, hoe kan dat? [online video] 19 February. Available at: www.rtl.nl/video/d0067dec-76a6-47c7-9c4c-042778be8ac6 (Accessed: 14 July 2020).

Jones, B., 2016. Authenticity in political discourse. *Ethical Theory and Moral Practice*, 19(2), pp. 489–504.

Kimmel, M., 2017. *Angry white men: American masculinity at the end of an era*. London: Hachette.

Leijten, J., 2019. Nederlandse boeren produceren grotendeels voor het buitenland. *NRC*, 5 October [online]. Available at: www.nrc.nl/nieuws/2019/10/05/nederlandse-boeren-produceren-grotendeels-voor-het-buitenland-a3975757 (Accessed: 14 July 2020).

Little, J., 2002. Rural geography: Rural gender identity and the performance of masculinity and femininity in the countryside. *Progress in Human Geography*, 26(5), pp. 665–670.

MacCannell, D., 1973. Staged authenticity: Arrangements of social space in tourist settings. *American Journal of Sociology*, 79(3), pp. 589–603.

Mamonova, N. and Franquesa, J., 2019. Populism, neoliberalism and agrarian movements in Europe: Understanding rural support for right-wing politics and looking for progressive solutions. *Sociologia Ruralis*, pp. 1–22.

Mascini, L., 2019. Aangifte tegen protesterende boeren op Mediapark. *De Gooi- en Eemlander*, 20 December [online]. Available at: www.gooieneemlander.nl/cnt/dmf20191220_96462412/aangifte-tegen-protesterende-boeren-op-mediapark-homofobie-en-racisme-horen-niet-thuis-bij-een-betoging?utm_source=google&utm_medium=organic (Accessed: 14 July 2020).

Molina, F. and Macho, A. M., 2016. The persistence of the rural idyll: Peasant imagery, social change and nationalism in Spain 1939–1978. *European Review of History: Revue Européenne D'histoire*, 23(4), pp. 686–706.

Nieuwsuur, 2019. Episode 256. 1 October [online]. Available at: www.npostart.nl/nieuwsuur/01-10-2019/VPWON_1303269 (Accessed: 14 July 2020).

NOS, 2019a. Boerenprotest Den Haag voorbij, 2200 trekkers terug naar huis. *NOS*, 1 October [online]. Available at: https://nos.nl/liveblog/2304125-boerenprotest-den-haag-voorbij-2200-trekkers-terug-naar-huis.html (Accessed: 14 July 2020).

NOS, 2019b. Deze boeren leggen uit waarom ze demonstreren. *NOS*, 1 October [online]. Available at: https://nos.nl/collectie/13799/artikel/2304207-deze-boeren-leggen-uit-waarom-ze-demonstreren-ik-zit-muurvast-door-dit-beleid (Accessed: 14 July 2020).

NU.nl, 2019. In totaal zeven aanhoudingen bij boerenprotest in Den Haag. *NU.nl*, 1 October [online]. Available at: www.nu.nl/binnenland/6001237/in-totaal-zeven-aanhoudingen-bij-boerenprotest-in-den-haag.html (Accessed: 14 July 2020).

Peeren, E., 2018. Villages gone wild: Death by rural idyll in *The Casual Vacancy* and *Glue*. In Stringer, B. (ed), *Rurality re-imagined: Villagers, farmers, wanderers, wild things*. Novato: Applied Research and Design Publishing, pp. 62–73.

Phvanmulligen, 2019. 18% van de boeren is miljonair. Gemiddeld hebben boeren besteedbaar inkomen van 42 duizend euro. -Van alle werkenden is 1,5% miljonair, en werkenden hebben een gemiddeld inkomen van 34 duizend euro [Twitter]. 1 October. Available at: https://twitter.com/phvmulligen/status/1179103445523062784 (Accessed: 14 July 2020).

Piotrowski, M., 2017. "Authentic" folds: environmental audiences, activists and subjectification in hypocrisy micropolitics. *Continuum*, 31(6), pp. 844–856.

RTL Nieuws, 2019. Sympathie voor boeren kalft af na nieuwe acties. *RTL Nieuws*, 18 December [online]. Available at: www.rtlnieuws.nl/nieuws/nederland/artikel/4960241/minder-steun-neemt-af-boeren-boerenacties-boerenprotest (Accessed: 14 July 2020).

RTL Nieuws, 2020. Boeren met trekkers toch op snelweg. *RTL Nieuws*, 19 February [online]. Available at: www.rtlnieuws.nl/nieuws/nederland/artikel/5027231/boeren-protest-tractor-trekker-snelweg (Accessed: 14 July 2020).

Ruitenbeek, A., 2019. Sieta van Keimpema militant? *Het Financieele Dagblad*, 15 October [online]. Available at: https://fd.nl/profiel/1319957/sieta-van-keimpema-militant-het-is-maar-wie-je-spreekt (Accessed: 14 July 2020).

Schoorl, J., 2019. Boerenactivisme? *De Volkskrant*, 20 December [online]. Available at: www.volkskrant.nl/nieuws-achtergrond/boerenactivisme-dat-gaat-bij-sieta-van-keimpema-nooit-op-stal-hoeveel-melk-er-ook-in-de-mestput-loopt~b393bbfa/ (Accessed: 14 July 2020).

Trilling, L., 2009. *Sincerity and authenticity*. Cambridge, MA: Harvard University Press.

Trouw, 2019. De staat van de boer. *Trouw* [online]. Available at: https://destaatvande-boer.trouw.nl/resultaten/ (Accessed: 14 July 2020).

Van Essen, J., 2019a. Frontman tegen wil en dank. *De Gelderlander*, 19 September [online]. Available at: www.gelderlander.nl/buren/frontman-tegen-wil-en-dank-ik-durf-mijn-dochter-amper-meer-aan-vriendinnetjes-te-laten-vertellen-dat-wij-schapen-houden~a1f1ec877/ (Accessed: 14 July 2020).

Van Velzen, J., 2019b. Wat wil de boer nou eigenlijk? *Trouw*, 29 December [online]. Available at: www.trouw.nl/economie/wat-wil-de-boer-nou-eigenlijk-vijf-inzichten-om-de-boerenprotesten-beter-te-snappen~bd9682dd/ (Accessed: 14 July 2020).

Varga, S., 2011. The paradox of authenticity. *Telos: Critical theory of the contemporary*, 156, pp. 113–130.

Verbeek, F., 2019. Tientallen aanhoudingen in Amsterdam. *Elsevier Weekblad*, 7 October [online]. Available at: www.elsevierweekblad.nl/nederland/achtergrond/2019/10/tientallen-aanhoudingen-op-stadhouderskade-werkloze-klimaathippies-713126/ (Accessed: 14 July 2020).

Verhoeven, N., 2019. Veevoederreus en slachthuisgigant betalen duizenden euro's aan boerenprotest. *De Gelderlander*, 2 October [online]. Available at: www.gelderlander.nl/home/veevoederreus-en-slachthuisgigant-betalen-duizenden-euro-s-aan-boerenprotest~ab590468/ (Accessed: 14 July 2020).

Winterman, P., 2019. Boerenprotest Groningen loopt uit de hand, woede in Tweede Kamer. *AD*, 14 October [online]. Available at: www.ad.nl/binnenland/boerenpro-test-groningen-loopt-uit-de-hand-woede-in-tweede-kamer~a02e96c1/ (Accessed: 14 July 2020).

Woods, M., 2006. Redefining the "rural question": The new "politics of the rural" and social policy. *Social Policy & Administration*, 40(6), pp. 579–595.

Zander, M. F., 2006. Musical influences in advertising: How music modifies first impressions of product endorsers and brands. *Psychology of Music*, 34(4), pp. 465–480.

Zwiers, M., 2015. Rebel rock: Lynyrd Skynyrd, Normaal, and regional identity. *Southern Cultures*, pp. 85–102.

9 Idyllic politics and politics of the idyll

Florian Dünckmann

9.1 Introduction: what is political about the rural idyll?

The idea of an idyllic countryside is a central topos of the imaginary geography of Western societies. According to this, rural life still runs in its natural circles, time passes more slowly, while people know their place in the community and live in close contact with nature. The social attributions contained in these representations of the rural – authenticity, stability and harmony – clearly bear traits of ideology. Therefore, a critical geography of the rural cannot avoid considering the political implications of such notions. Indeed, many authors have already dealt with the historical roots and ideological patterns of idyllic images and tried to unmask their falseness (e.g. Marx 1964, Williams 1973, Bunce 1994). In this chapter I want to add a new dimension to the discussion by taking a fresh look at the political content of the idyll. In doing this, my central argument is this: Although the image of the idyll in itself seems to be a thoroughly un- or anti-political concept, it still – or precisely because of this – has a strong argumentative power in political debates. They revolve around whether and where idyllic ideas can or should have a legitimate place in our society. In other words, there is a "politics of the idyll" – that is, a political debate about the meaning that the desire for harmony, stability and authenticity should have in our world.

I draw primarily on the political philosophy of Hannah Arendt, whose writings have dealt in depth with the relationship between private and public and with the social dangers of those "anti-political" ideologies that seek to counter the complexity of political debates with an emphasis on simplicity and naturalness. In doing so, Arendt has always defended the place of the political – that is, the contentious confrontation of different concepts of life and world, in our society. Based on her concept of politics, I distinguish between idyllic politics –the employment of political utopias of idyllic coexistence – on the one hand, and the political effectiveness of the idyll and the anti-idyll on the other. Using two examples – the personal memories of a World War II refugee from former East Prussia, and the political arguments of conventional farmers against a greener agricultural policy – I want to show in what different areas and in what different ways ideas of the idyll can be effective in political debates.

DOI: 10.4324/9781003091714-11

9.2 Idyll and anti-idyll

Many who had their childhood and youth in the 1970s will remember the television show *The Waltons*. Its focus is on the extended family of the Waltons during the US economic crisis of the 1930s. The family leads a poor, God-fearing life full of privations, but also full of human warmth and family cohesion in rural Virginia. The opening credits of the series show the family receiving their first radio and gathering around their new acquisition, for which they have had to save their money for a long time. At the end of each episode, viewers see the lights of the Waltons' peaceful home going out and hear the family members wishing each other a good night. In short, the Waltons embodied a rural idyll through and through. Yet the picture painted by the Waltons in this TV show radically contradicts the actual living conditions that the poor, marginalised rural population in the eastern United States – the so-called hillbillies – endured at the time of the economic crisis. The Waltons knew no domestic violence, no family abuse, no overt racism and no rampant tooth decay. Compared with historical reality, the Waltons appear to be one big lie – not only being untrue, but also serving a clear ideological purpose. When Chopra-Gant (2013) compares the Waltons to a political narcotic, he is describing a function that many authors attribute to different notions of the idyll.

In his classic study for Great Britain, Williams (1973) traces the development of the modern idea of the rural idyll in literature and art and concludes that it has its origins in the phase of industrialisation and urbanisation. During this transition from an agrarian to an industrial society, the former complex social structures of rural communities were increasingly pushed into the background, while their retrospect representation simultaneously transfigured and simplified them. The origin of the notion of the rural idyll is thus to be found in an urban environment that looks at the countryside from outside. It is this bourgeois perspective, with its penchant for nationalism, that attributes values such as social stability, authenticity and naturalness to the countryside (Bunce 1994). Moreover, the image of a rural idyll also shaped concrete political ideas and planning activities: Model rural settlements emerged that were informed by social-utopian models of a stable, hierarchically ordered social structure (Havinden 1989). At the beginning of the 20th century, social reform movements were also gaining ground throughout Western Europe, propagating a return to the countryside and a natural lifestyle. Numerous authors have pointed out how much this idealised picture of the rural excludes marginalised groups and cleanses the image of the countryside of all existing social ruptures and contradictions (e.g. Keith 1989; Cronon 1991; Short 1991; Bell 2006). And even in today's late-capitalist society, this longing for a rural primeval state, which our society has supposedly lost in the course of its increasing urbanization, manifests itself in very different social phenomena. In his detailed analysis of various past and present 'idylls', Baumann (2018) identifies discourse elements of idyllic rural life in such diverse phenomena as periurbanisation, *urban gardening* and the boom in rural magazines.

However, this represents only one side of the multifaceted discourse. Alongside the rural idyll a complementary figure represents a pattern of thought that is at least as deeply rooted and influential. This is the idea of the anti-idyll – that is, the false idyll, which has to be unveiled by a supposedly realistic glimpse behind the harmonious façade where cold social mechanisms of power, injustice or even violence hide from public view. Complementing those idyllic hillbillies the Waltons, Bell (1997) identifies the topos of the so-called Hillbilly Horror in films such as *Texas Chainsaw Massacre* and *Friday the 13th*. Its narrative structure is always similar: A group of young people from the city become stranded in a remote rural area and fall victim to the murderous lust of the degenerated, bloodthirsty country dwellers. Like the rural idyll, this is an outside view of the countryside, with its supposedly hidden reality now being a projection of modern fears instead of urban longings. More subtle, but in their discursive structure quite similar representations of a country life, in which chill and terror is hidden behind an idyllic façade, can also be found in numerous films (e.g. Michael Hanekes' *The White Ribbon*) or in literature (e.g. in the works of Thomas Bernhard) and the fine arts (e.g. in pictures by Neo Rauch). In summary, in modern society the image of the countryside fluctuates between the poles of idyll and anti-idyll, both being expressions of an outside perspective.

Against this background, the question of whether and to what extent the rural idyll actually exists, or whether it is a lie behind which an anti-idyll is hidden, appears not to be productive at all. Furthermore, I believe that most of the readers of this chapter, even though they might fully acknowledge the discursive, historically contingent and ideological character of any social notion of the idyll, have nevertheless experienced moments and places that they would actually describe as idyllic. Given this apparent contradiction, I believe it to be more productive to grasp the phenomenon of idyllic notions not with the categories of truth or lie, but rather with the concept of kitsch.

9.3 Kitsch or the second tear

Cultural studies and art theory have long dealt with the phenomenon of kitsch in many different ways (cf., e.g., Dorfles 1969; Călinescu 1987; Kulka 1988). They show how this seemingly banal category can provide deep and valuable insights into the cultural, social and political conditions of late modernity and, to an extent, into the meaning of the rural idyll and rurality in general.

For a start, it is clear that the accusation of kitsch does not relate to the truth content of a (mostly artistic) representation, but rather refers to its aesthetics, appropriateness or situational classification. Take the (fictional) example of a documentary film that claims to make true statements: A closing shot showing a sunset is neither true nor false; however, one can consider it as insincere, sentimental, out of place or kitschy. Originally meant as a pejorative expression for unoriginal, cheaply mass-produced imitations of luxury goods, kitsch was perceived as a mirror of the aesthetic preferences of the lower classes and became a marker for their bad taste. Critics such as Adorno and Horkheimer (1944[1977]) and Broch (1969) saw it simply as a manifestation of the vulgarity of capitalist mass culture with its tendency to turn everything valuable into a marketable consumer object.

However, in recent years increasing numbers of authors have called for a more nuanced exploration of the phenomenon of kitsch, and have tried to understand its conditions and mechanisms (e.g. Pieterse 2002; Atkinson 2007; Irazabal 2007; Hwang and Ling 2008; Mukhopadhyay 2008;). In his seminal paper, Binkley (2000) presented kitsch as an aesthetic strategy that deals with the modern experience of ontological insecurity. Kitsch

> glories in its embeddedness in routines, its faithfulness to conventions, and its rootedness in the modest cadence of daily life, works to re-embed its consumers, to replenish stocks of ontological security, and to shore up a sense of cosmic coherence in an unstable world of challenge, innovation and creativity.
>
> (Binkley 2000, 135)

Kitsch's aesthetic clumsiness and lack of innovation should signal humanity and universality. Most importantly, in its emphatic sentimentality – its "feeling for feeling" (142) – it claims to point to basic, authentic and universal human conditions. In this sense, the aesthetics of kitsch show remarkable parallels with the idea of the idyll in that they both celebrate simplicity, authenticity and the mundane.

However, with all its asserted innocence, kitsch can turn out to be a politically questionable form of representing reality. In addressing the "real universality of modest origins, of common fellowships between all people, of love for that which is commonly, undeniably and obviously lovable" (Binkley 2000, 146) it is susceptible to anti-elitist and populist ideas. Therefore, it is not surprising that totalitarian regimes often tend towards kitschy aesthetics when appealing to the imagined community of a nation or a people. Stalin made ample use of it (Bullock 2006, Franz 2007), and the Nazis tried to monopolise their own kitschy political symbolism by waging a so-called "campaign against kitsch" (Skradol 2011).

In his novel *The Unbearable Lightness of Being*, Milan Kundera gives a remarkable definition of kitsch. Among other things, the novel deals with the question of what significance the search for private happiness can have against the background of political events such as the suppression of the Prague Spring. In this context, Kundera puts the following statement into the mouth of Sabina – the character of a serene, urban artist:

> Kitsch causes two tears to flow in quick succession. The first tear says: 'How nice to see the children running on the grass!' The second tear says: 'How nice to be moved, together with all mankind, by children running on the grass!' Only this second tear makes kitsch kitsch.
>
> (Kundera 1984, 251)

According to this definition, the insincerity of kitsch consists less in the original feeling of being touched, and more in the fact that it carries these immediate impressions, emotions and affects out of the realm of private and individual intimacy, transforms them into models of a better world, and thus robs them of their strict locality, temporality and situativity. Thus, if there are truly idyllic moments and places, they can only exist as long as they only stand for themselves. As soon as they are pointed out and represented, named and given meaning – in other words, as soon as the "second tear" flows – they change their character and become kitschy, inappropriate or sentimental. Accordingly, the idyll, if at all, can only exist in the private sphere and is not compatible with the collective view of the public.

This idea of a fundamental difference between the private and the public sphere is also the basis of Hannah Arendt's political philosophy, as expressed in *The Human Condition* (Arendt 1958 [1998]). Although feminist theory in particular has criticised her (with good reason) for this strict dichotomy (cf. e.g. Honig 1995; Benhabib 1996), the next section will show to what extent Arendt's definition of the political, which is fed by precisely this juxtaposition of private and public space, can prove productive for understanding the politics of the idyll.

9.4 Hannah Arendt: defending disunity against harmony

Hannah Arendt's philosophy is largely devoted to the defence of the political – that is, that area in which the plurality of people, their diversity and their different perspectives can find expression. Her personal experience of Fascism and her attempt to understand the origin and logic of totalitarian systems, whose central concern is to bring the realm of the political and dynamic civil society into line and thus suffocate them, have always shaped her way of thinking. In her view, politics is the active handling of the multi-perspectivity and disunity that result from the fundamental diversity of people. For Arendt, the venue and subject of every political debate is the *world* – that is, the space of public interaction that we are forced to share with those neither we nor they have chosen as our neighbours:

> To live together in the world means essentially that a world of things is between those who have it in common, as a table is located between those who sit around it; the world, like every in-between, relates and separates men at the same time.
>
> (Arendt 1958, 52)

Arendt contrasts this open, non-exclusive *world* with the private realm. This is the area in which the basic necessities of human life and well-being are organised and which should be characterised by personal attention, human warmth and existential security. Arendt insists on the clear separation of these two areas, where different rules of coexistence and different definitions of right and wrong would apply. In the private sphere, political-public values such as equality and power symmetry have no place. On the other hand, values that actually have a positive connotation, such as love or goodness, belong exclusively in the private sphere and should not be used as guidelines for politics. In contrast to friendship or solidarity, for example, it is a characteristic of love that it always refers to very specific, non-substitutable people; it is exclusive and thus not part of the space of public interaction. This is why "love can only become false and perverted when it is used for political purposes such as the change or salvation of the world" (Arendt 1958, 52).

Even if Arendt never writes it, she sees the public-political space primarily as an urban space; it is no coincidence that the word "politics" is derived from the ancient city-state, the polis, which Arendt sees as paradigmatic for the western conception of politics (Arendt 1958, 201). In contrast, the sociologist Ferdinand Tönnies (1887) painted a picture of the rural village whose social structure consists of an exclusive, coherent community being familiar to all its members. Such a village would represent the exact antithesis to Arendt's *world*, which is characterised by openness, diversity and disunity (Arendt 1958, 50–58). Thus, above all, Arendt's philosophy is a warning against all utopias that – either by purification or education – aim to transform society into an oversized, harmonious, village-like community. By ultimately seeking to abolish or overcome political differences, those attempts always promote totalitarian orders.

According to Arendt, fundamentally positive values such as harmony, love, community, goodness and naturalness belong in the private sphere and can have disastrous and dangerous effects if one transfers them to the political sphere of the public. This obviously also applies for the subject of this article, the idyll: It is precisely this inappropriate transfer of private values into public values, as we have seen, which causes unease in Arendt. Therefore, political kitsch, whose evocation of idyllic conditions feed the 'second tear', is not only inappropriate or sentimental, it is also politically dangerous because it implicitly appeals to overcome all political differences in the name of something more fundamental than political conflict, be it God, nature, race or country. To those who set this ideal above all political disagreement, the idea itself suggests that the political community must then be purified of all those who stand in the way of this ideal and who can thus no longer be part of that community.

Such an evocation of the idyll as the basis of a deeply anti-democratic and racist ideology is found in Germany, for example, among the so-called *Völkische Siedler* – that is, right-wing extremist drop-outs who have turned their backs on the supposedly depraved city life and have settled in rural areas. They seek to implement and propagate their idea of an independent, "ethnically pure" community that cultivates its customs and lives in harmony with the earth. Many of these people see themselves as a continuation of the so-called Artamans, an anti-Semitic back-to-the-countryside-movement of the 1920s. They practise traditional arts and crafts or organic farming and usually act as friendly nature- and family-loving neighbours in the villages where they settle. At the same time, however, they are firmly embedded in right-wing extremist networks (e.g. national youth organisations) and see themselves as defenders of a biologically 'pure' national community (Schmidt 2014).

Their explicit reference to the agrarian romanticism of the National Socialists shows these "racial-radical idylls" (Baumann 2018, 103) as extreme examples of a political kitsch that makes use of the evocation of a supposedly authentic past being lost in the course of modernisation, globalisation and alienation. However, political kitsch in the form of idyllic utopias has many forms and can also be observed in completely different contexts: Election ads of political parties that depict the everyday life of content people in green landscapes; left-wing politicians, such as Evo Morales in Bolivia, who make use of a restorative discourse and base their political concept on the desire to regain the original authentic lifestyle that has been corrupted by the influence of the West (cf. e.g. Fladvad 2017). Such political thinking is what Žižek (2000, 190) calls "Arche Politics": it turns away from the world, which according to Arendt is characterised by plurality and multi-perspectivity, and conjures up a vision of authenticity and naturalness that is supposedly beyond all disputes and political disagreement. One could say that in this vision of an original, homogenous and organic social space the political has no place, since everything is 'in its place'.

9.5 Both sides of the garden fence: politics of the idyll

At this point, we could end this train of thought and conclude with Arendt that the idyll can only exist in the immediacy of the private sphere and that, accordingly, it is out of place in the public sphere. However, in my opinion the connection between the idyll and politics turns out to be much more complex. The reason for this is the fact that individual experiences of the idyll are always relational – that is, they acquire their meaning only through their relationship to other experiences and moments. It is therefore not surprising that the concept of idyllic country life originates from the era of industrialisation and urbanisation and that country magazines celebrate a simple lifestyle at the same time that life in our digitalised, late-modern society is becoming increasingly complex. Figuratively speaking, my private garden idyll only acquires its meaning in the context of a surrounding world that I perceive as non-idyllic. Last but not least, the dividing line between the private and the political is itself the subject of political debate. Perhaps there can be agreement that my private garden idyll should be separated from the public sphere. However, the question of where the boundaries of my garden are – that is, where my garden fence runs and to what extent things happening there also legitimately concern people outside – is clearly a public question in which plurality, multi-perspectivity and disagreement play an important role.

From this dialectic of private and public – that is, the fact that the private and public spheres are only formed by their demarcation from each other – emerges a politics of the idyll that controversially negotiates questions about the legitimacy of idyllic ideas. On closer inspection, it becomes clear that no small part of the political debates in and around rural issues show characteristics of such a politics of the idyll. Two cases show how different these debates can be. In fact, these two examples were deliberately chosen for their dissimilarity in order to mark opposite sides in the wide spectrum of the politics of the idyll – that is, the different ways in which ideas of a rural idyll are evoked, rejected, questioned and legitimised in political debates.

The first example refers to a bachelor's thesis (Hanson 2015) in which a student investigated the imaginary geography of her grandmother, whose identity is largely nourished by her memories of the East Prussian village of her childhood and youth, from which she had to flee at the end of World War II at the age of 16. In poems, oral tales and reports, with self-painted maps and collected objects of local craftsmanship, she still conjures up her lost homeland. The following excerpt from a poem reads like the ideal image of a rural idyll:

> Der Wind raschelnd durch die Pappeln blies/Es war fürwahr ein kleines Paradies/Lerchen trillerten ihr Lied hoch in der Luft/Und die Birken hatten den herrlichen Duft/Vom hohen Dache, horch, horch (…)/Da klapperte der Storch/Und friedlich die Kühe auf der Wiese weiden/Leider mussten wir von all dem scheiden/Ein Jeder von uns an die schöne Heimat denkt/Sie war glanzvoll, die hatte uns der Herrgott geschenkt.
>
> [The wind rustling through the aspens/It was a little paradise indeed/ Larks trilled their song high in the air/And the birches had the wonderful smell/From the high roof, listen, listen (…)/There the stork clattered / And the cows grazed peacefully on the meadow/We had to part from all this/Each of us thinks of the beautiful homeland/That was splendid, what the Lord God had given to us.]
>
> (Hanson 2015, 20)

With regard to a politics of the idyll, a multi-layered picture emerges: First, beyond all political instrumentalisation, this poem can be regarded as the expression of a very personal construction of memory. In an attempt to give her own biography a coherent structure and meaning, the concrete East Prussian village represents the past anchor point of an individual, irreplaceable life story, which can be schematically summarised as a "displacement from the idyll". However, since experiences of idyll are always relational, the village is only perceived as idyllic through the lens of later experiences of loss; in a sense, paradise only exists as a result of the expulsion from it. And so, even though there is talk of aspen, larks, storks and cows, the violent history of the 20th century remains present throughout this idyllic poem.

Moreover, the individual stories of flight and expulsion in post-war Germany were never "innocent", since they were always the subject of identity politics and competition for memory. In this way, they were used by the *Vertriebenenverbände* – that is, the associations of expellees from Germany's former eastern provinces – for revanchist demands or even for relativising the crimes of National Socialism (Picard 2017). Others questioned the validity of any of these laments over personal losses with regard to these same crimes of National Socialism (Kossert 2009). The personal biographies of this generation, whose lives were so closely entangled with the European experiences of violence in the 20th century, are thus condemned to be embedded in a political context in which personal memories of idyllic moments are always more than just private matters. At the same time, however, it falls short of the mark if we deny them any authenticity and merely regard them as political kitsch, to be used as a strategic instrument in 20th-century memory wars (Rothberg 2009).

The second example of a policy of the idyll is the political debate on the transition to a more ecologically sound agriculture, as demanded by the German Green Party and numerous environmental groups. Against this demand, the representatives of conventional agriculture, and primarily the farmers' association, put forward the following argument: The model of an environmentally friendly, animal welfare-conscious and farm-based agriculture is an outdated and unrealistic idea of agricultural production that ignores the real structures and requirements of modern animal and plant production. Proponents of conventional agriculture called such ecological ideas "Bullerby-Politics", referring to Astrid Lindgren's popular children's books depicting an idyllic childhood in Bullerby, a fictive village in the Swedish Countryside. Thereby they clearly refer to the ideological figure of the false idyll or the anti-idyll. According to this, green ideas about agriculture are suspect and ultimately wrong for the very reason that they are idyllic. Behind the idyllic veil, however, there is no injustice and violence in this case, but rather economic constraints and unsentimental business considerations.

In this way, the accusation of following idyllic ideas is primarily an effective rhetorical figure in a concrete political conflict of interests. On a more fundamental level, however, it is also about the political question of the legitimacy of – in Arendt's sense – "alien" ideas and motives for action in the public sphere of politics. On one side, there are accusations of a retreat into the private sphere, of being cut off from the world and thus of a self-referential bourgeoisie. On the other side, there is the argument that the longing for authenticity, stability, harmony and a private space are legitimate desires and thus constitute an important component of a "good life", the facilitation of which should ultimately be the yardstick of any political debate. Is it a legitimate political demand, then, for people to want agriculture that matches their idyllic image? Or does its idyllic nature alone disqualify it as a guideline for practical politics? The following quote from an interview with the politician Robert Habeck from the Green Party illustrates this ambivalence between the desire for authenticity or the "good life" on the one hand, and the desire not to talk political kitsch on the other:

"There's nothing wrong with Bullerby. And Astrid Lindgren was a social revolutionary. Every Green party member can refer to her. And thus also to an idyll: the picture of the house with garden, buttercups, swing in the apple tree and robber's daughter nights. The hedges don't have to be cut at right angles, the house may be a little crooked, and there may be a few worms living in the apples".

(Interview Frankfurter Allgemeine Zeitung, from 07.10.2013, Translation FD)

If the idea of the idyll is an important element in the way modern individuals perceive and experience their relationship to the world, then it is also obvious that it becomes an important object of public debate about this world and thus of politics. The question of the legitimate place that notions of authenticity, harmony and stability should have in a democratic, pluralistic and just society is always under discussion. The two examples presented here show in what different contexts and in what different ways this question becomes relevant.

9.6 Conclusion

In an unequal society, even the ideals of the idyll – whether Bullerby or the Waltons – depend on class and lifestyle. What all these idylls have in common, however, is that they project their desires onto the rural area and construct it as a potentially apolitical space.

In *The Unbearable Lightness of Being*, two of the main characters, Teresa and Tomas, finally move to a Bohemian village, being exhausted and disappointed by the political events following the Prague Spring. Teresa, in particular, sees rural life as a melancholic and purified alternative to the failed political commitment that has marked her life in the city. Many people who move to the countryside imagine the village as a place where unquestioned self-evidence and constancy still prevail and which thus offers relief from the permanent challenge of critical reflection and political debate (Duenckmann 2010). A politics of the idyll – that is, the debate on the question of what position and what significance such ideas should have in public space – runs through many debates in rural areas, whether it be the expansion of renewable energies, dealing with displaced persons or the role of agriculture. Any critical study of rural areas that deals with social inequalities, economic exclusion and expropriation processes, everyday discrimination, and visible or invisible practices of resistance must therefore also deal with the political dialectics of the idyll: The idyll is not only the idea of a place that is excluded from political debates, but also the object and subject of political debates.

Bibliography

Adorno, T. and Horkheimer, M. (1944[1977]) *Dialektik der Aufklärung. Philosophische Fragmente*. Frankfurt A.M.: Fischer Verlag.

Arendt, H. (1958) *The human condition*. Chicago/London: University of Chicago Press.

Atkinson, D. (2007) "Kitsch geographies and the everyday spaces of social memory", *Environment and Planning A*, 39 (3), pp. 521–540.

Baumann, C. (2018) *Idyllische Ländlichkeit. Eine Kulturgeographie der Landlust*. Bielefeld: Transcript.

Bell, D. (1997) "Anti-idyll: Rural horror" in Cloke, P., Little, J. (eds.): *Contested countryside cultures – otherness, marginalisation and rurality*. London/New York: Routledge, pp. 94–108.

Bell, D. (2006) "Variations on the rural idyll" in Cloke, P., Marsden, T., Mooney, E. (eds.) *The handbook of rural studies*. Los Angeles: SAGE, pp. 149–160.

Benhabib, S. (1996) *The reluctant modernism of Hannah Arendt*. Thousand Oaks: SAGE.

Binkley, S. (2000) "Kitsch as a repetitive system", *Journal of Material Culture* 5 (2), pp. 131–152.

Broch, H. (1969) "Notes on the problem of Kitsch" in Dorfles, G. (ed.) *Kitsch: The world of bad taste*. New York: Universe Books, pp. 49–76.

Bullock, P. (2006) "Andrei Platonov's 'Happy Moscow' Stalinist kitsch and ethical decadence", *Modern Language Review* 101 (1), pp. 201–211.

Bunce, M. F. (1994) *The countryside ideal. Anglo-American images of landscape.* London/ New York: Routledge.

Călinescu, M. (1987) *Five faces of modernity. Modernism, avant-garde, decadence, kitsch, postmodernism.* Durham: Duke University Press.

Chopra-Gant, M. (2013) *The Waltons. Nostalgia and myth in seventies America.* London/ New York: I.B.Tauris.

Cronon, W. (1991) *Nature's metropolis. Chicago and the great west.* New York: Norton.

Dorfles, G. (ed.) (1969) *Kitsch. An anthology of bad taste.* New York: Universe Books.

Duenckmann, F. (2010) "The village in the mind: Applying Q-methodology to reconstructing constructions of rurality", *Journal of Rural Studies*, 26 (3), pp. 284–295.

Fladvad, B. (2017) *Topologien der Gerechtigkeit. Eine politisch-geographische Perspektive auf das Recht auf Ernährungssouveränität in Bolivien.* Department of Geography, Kiel University.

Franz, N. P. (2007) "Stalin kitsch? Remarks on kitsch, on Ciaureli's Film 'Kljatva' (1949) and on German's Rasskazy o pervom cekiste (1946)", *Zeitschrift für Slawistik*, 52 (4), pp. 447–474.

Hanson, Liv-Lareen (2015) *Erinnerung als Tätigkeit: Individuelle Erinnerungspraktiken einer Heimatvertriebenen.* Kiel: Bachelor Thesis at the Department of Geography, Kiel University.

Havinden, R. (1989) "The model village", in Mingay, G. E. (ed.) *The rural idyll.* London: Routledge, pp. 23–36.

Honig, B. (Ed.) (1995) *Feminist interpretations of Hannah Arendt.* University Park, PA: Pennsylvania State University Press.

Hwang, C.-C. and Ling, L. H. M. (2008) "The 'Kitsch' of war misappropriating Sun Tzu for an American imperial hypermasculinity", *International Affairs Working Paper*, 2008–04, pp. 1–23.

Irazabal, C. (2007) "Kitsch is dead, long live kitsch: The production of hyperkitsch in Las Vegas", *Journal of Architectural and Planning Research*, 24 (3), pp. 199–223.

Keith, W. J. (1989) "The land in Victorian literature", in Mingay, G. E. (ed.) *The rural idyll.* London: Routledge, pp. 77–90.

Kossert, A. (2009) *Kalte Heimat. Die Geschichte der Deutschen Vertriebenen Nach 1945.* München: Pantheon.

Kulka, T. (1988) "Kitsch", *British Journal of Aesthetics*, 28 (1), pp. 18–27.

Kundera, Milan (1984) *The Unbearable Lightness of Being.* London: Faber and Faber.

Marx, Leo (1964) *The Machine in the Garden: Technology and the Pastoral Ideal in America.* New York: Oxford University Press.

Mukhopadhyay, B. (2008) "Dream kitsch-folk art, indigenous media and '9/11': The work of pat in the era of electronic transmission", *Journal of Rural Studies*, 13 (1), pp. 5–34.

Picard, L. (2017) "Tout est politique!" Eignet sich die politische Komponente für die Bewertung eines schlesischen Heimatblatts? in Kasten, T., Fendl, E. (eds.) *Heimatzeitschriften. Funktionen, Netzwerke, Quellenwert*. Münster/New York: Waxmann, pp. 77–94.

Pieterse, J. N. (2002) "Globalization, Kitsch and conflict: Technologies of work, war and politics", *Review of International Political Economy*, 9 (1), pp. 1–36.

Rothberg, M. (2009) *Multidirectional memory. Remembering the holocaust in the age of decolonization*. Stanford: Stanford University Press.

Schmidt, A. (2014) *Völkische Siedler/innen im ländlichen Raum. Basiswissen und Handlungsstrategien*. Berlin: Amadeu Antonio Stiftung.

Short, J. R. (1991) *Imagined country. Environment, culture, and society*. London/New York: Routledge.

Skradol, N. (2011) "Fascism and Kitsch: The Nazi campaign against Kitsch", *German Studies Review*, 34 (3), pp. 595–612.

Tönnies, F. (1887) *Gemeinschaft und Gesellschaft. Abhandlung des Communismus und des Socialismus als empirischer Culturformen*. Leipzig: Fues.

Williams, R. (1973) *The country and the city*. London: Hoghart Press.

Žižek, S. (2000) *The Ticklish subject. The absent center of political ontology*. New York: Verso.

10 Dystopia as authenticity

Changing ruralities in Icelandic cinema

Thoroddur Bjarnason, Brynhildur Thorarinsdottir and Menelaos Gkartzios

Introduction

The film industry is a predominantly urban and urbane enterprise. Directors, script-writers, actors and other artists and technicians directly involved in film-making are likely to live and work in cities, as are the people involved in financing, producing and promoting films. The majority of the audience generally also lives in urban areas, and the urban audience may loom even larger in the minds of those involved in the film production process (Pratt 2007). The urban gaze of cinema may take various forms, including, for example, nostalgic accounts of rural childhoods, hyper-idyllic portrayals of small-town Christmases or decidedly negative stories of rural idiocy, backwardness, isolation and decline. As such, cinema creates and reinforces the conditions of commodification, tapping into dominant and predominantly urban constructions of rurality.

Feature films reflect the taken-for-granted social reality of those involved in its production and their assumptions about their intended audiences. More importantly, such cinematic storytelling contributes to the constant repetition of "self-evident truths" that actively maintain and shape the economy of social power and channel social action, with profound political, economic and social consequences (Foucault 1970, 1978; see also Nerlich and Döring 2005; Brandth and Haugen 2011). The emotional and cognitive cinema experience of *having been there* can be expected to shape and reinforce ideas about people, places and events, in particular for those with no direct personal experience of the subject matter (Sutherland and Feltey 2013). This, in turn, may have wide-ranging effects on the politically viable policies for dealing with various social issues.

As critical theorists have established (e.g. Weber 1946; Marcuse 1964; Gramsci 1971; Foucault 1977), political authority is ultimately grounded in perceived legitimacy. Politically viable policies for dealing with, for example, corporate crime, international conflict, pandemic threats, street crime, political corruption or rural marginalisation must thus be consistent with the dominant worldview in society. However, few people have direct experiences with the boardroom of corporate headquarters, the hardship of common soldiers in a warzone, the outbreak of a deadly plague, organised crime and

DOI: 10.4324/9781003091714-12

political corruption, or January morning in an isolated fishing village just beyond the Arctic Circle.

While academic research, newscasts, documentaries and other forms of non-fiction are important sources of empirical facts, such facts are invariably interpreted within a framework of self-evident truths, moral conceptions, perceived risks and opportunities, and societal definitions of in-group and out-group (Bartlett 1923; Piaget 1932; Berger and Luckmann 1966). Similarly, the impressions of other places gained from travelling to distant countries, across the railroad tracks to the shadier part of town, or indeed out to the countryside are interpreted within such an overarching worldview with specific assumptions about, for example, race, gender and class, as well as the fundamental meaning of various other social categories, including the distinction between urban and rural. The vicarious experiences of social life provided by feature films are thus an important part of the taken-for-granted realities of a complex, global world.

According to Hechter's (1975) concept of internal colonialism, a strong contrast between the sophisticated, modern centre and the rough, underdeveloped peripheries is important in building and maintaining the nation-state. Rather than regional differences slowly fading away under the force of modernisation, images of such differences are carefully nurtured and exaggerated in order to maintain the cultural, political and economic hegemony of the centre over the peripheries. Prior studies have shown the role of feature films in defining the racist, violent and poor "others" in the US south (Jansson 1995) and their xenophobic, racist, sexist, homophobic and welfare-dependent counterparts in Sweden's north (Eriksson 2010) in contrast to the tolerant, modern and progressive urban mainstream. In the process, structural inequalities of class and rurality are recast as personal failings based on individual choices of lifestyle, culture and (im)mobility.

As a small, peripheral island nation in the middle of the Atlantic Ocean, Icelanders have certainly been subject to various stereotypes of the "other", including, for example, strong, bearded men and beautiful, promiscuous women in a rugged landscape and foul weather; an infestation of elves, trolls and ghosts; heavy consumption of strong alcohol; and a questionable cuisine of ram testicles, rotten shark, sour whale meat and boiled sheep-heads (see, e.g., Fontaine 2006; Alessio and Jóhannsdóttir 2011; Huijbens 2011). Iceland also has a surprisingly vibrant film industry that has effectively marketed the perceived quirkiness of Icelandic rurality among an international audience, yielding more than 400 awards at international film festivals in the period 2015–2019 (Icelandic Film Centre 2020a).

This chapter explores the portrayal of rurality in Icelandic film that manifests and shapes the dominant urban worldview of policymakers with profound real-life implications for rural Icelanders. This includes several interrelated themes, such as the razor-sharp distinction between urban and rural; representations of backwardness, decay and despair in rural Iceland against exaggerated depictions of the metropolitan swagger and the supposedly mean streets of the small, peripheral city of Reykjavík; toxic masculinity as the fundamental root of rural problems; and the perceived inevitability of unidirectional rural-to-urban migration. Such dystopic representations of rural Iceland have been successfully commodified and marketed internationally as "rural authenticity" to the critical acclaim of independent film-makers and their audiences, in turn reinforcing the urban hierarchy and providing support to the strongly urban-centred social and regional policies of Iceland.

The film industry and social development in Iceland

After centuries of Norwegian and later Danish rule, Iceland became a sovereign country under the Danish king in 1918. At that time, the entire population of the country was less than 92,000 (Statistics Iceland 2020a). The capital of Reykjavík was by far the largest town in the country, with about 15,000 inhabitants. Although the importance of traditional sheep and cow farming was on the decline, the farming communities still accounted for more than half the national population.

Of the 17 Icelandic feature films produced between 1918 and the establishment of the Icelandic Film Fund in 1978, 11 were set in rural Iceland and an additional three films had a strong rural dimension. From the outset, reviews in the local media reflected a preoccupation with the portrayal of Iceland to international audiences. A century later, Icelandic films based on the carefully constructed "authenticity" of rural Iceland have become an export commodity, supporting a vibrant film industry in a country with only 364,000 inhabitants.

Victor Sjöström's (1918) *Berg-Ejvind och hans Hustru* (*The Outlaw and his Wife*) followed sheep-thieving outlaws and lovers on the run from social injustice and corrupt authorities in the 18th-century Icelandic highlands. The film was well received abroad when it premiered simultaneously in Sweden, Denmark and Norway on New Year's Day 1918 and is considered to be a classic of the silent film era (Forslund 1988). The local Icelandic newspapers, however, lamented the way Iceland was misrepresented to an international audience, including Icelandic nature, traditional housing styles and traditional dress, and in particular how ugly the foreign sheep looked in comparison to their lovely Icelandic counterparts (Landið 1918; Fram 1920). A scathing contemporary newspaper review bewailed

No, there is surely no greater source of false information than irresponsible movies. … It is now shown all over the world and – of course – everyone who does not know better thinks this is a true picture of Icelandic people, Icelandic nature and Icelandic farms and dress!

(Fram 1920)

During World War II, British and later US forces occupied Iceland in order to secure shipping routes between North America and Britain. In 1944, the country declared full independence from Nazi-occupied Denmark and the Republic of Iceland was established. The population of the country was now about 128,000 and Reykjavík had swelled to 44,000. The population of the farming communities, however, had decreased by more than a quarter since 1918 and accounted for less than 30% of the population. Traditional farming culture was nevertheless still central to Icelandic national identity, as exemplified by the first Icelandic film in sound and colour: Loftur Guðmundsson's (1949) *Milli Fjalls og Fjöru* (*Between Mountain and Shore*), about a young farmboy wrongly accused of stealing sheep. The film raised fierce debate in Icelandic newspapers about whether the Icelandic landscape was more beautiful in colour or in black-and-white (Alþýðublaðið 1949; Útvarpstíðindi 1949), although the beauty of Icelandic women was thought to be well represented. Somewhat prophetically, one Icelandic newspaper declared that "this film raises hopes that Icelandic movies can in the future become an export commodity" (Alþýðublaðið 1949).

The growth of the city of Reykjavík and the influences of the American military base in the once rural Icelandic society was the underlying theme of Danish director Erik Balling's (1962) feature film *79 af Stöðinni* (*The Girl Gogo*). The protagonist is a young taxi driver who has moved from his family farm to the growing city. Through his line of work, he discovers the slimy underbelly of the city, which culminates in the realisation that his mysterious, urbane girlfriend funds her lavish lifestyle by receiving visiting officers from the American army base. Heartbroken, he tries to flee the city, but counter-urbanisation is not an option and after, falling asleep at the wheel, he ends up dead in a ditch. While generally pleased with the technical and artistic quality of the film, local critics nevertheless worried about its ability to thrust Iceland onto the international film stage: "Icelandic moviegoers will no doubt enjoy the movie. It is another question if it has the power of attraction necessary to become popular among the millions abroad" (Dagur 1962).

In 1978, parliament established the Icelandic Film Fund in order to support the production of domestic films. The population of Iceland had now reached 224,000 and the Reykjavík Capital Area had established a comfortable 53% majority within the country, while the farming communities dwindled to less than 25,000, accounting for only 11% of the population. Two of the three films supported by the Icelandic Film Fund provided dramatic accounts of young men struggling to break their rural bonds and move from their respective family farms to the city. In an emotionally charged and gushing review of Ágúst Guðmundsson's (1980) *Land og Synir* (*Land and Sons*), the film critic of Iceland's largest newspaper declared: "And now it is spring in Iceland, no less than a new form of art is being born. ... The age of cinema has arrived in our country" (Morgunblaðið 1980). Since 1980, Icelandic films have been released every year, growing from an average of 3.0 per year in 1980–1989 to 7.8 per year in 2010–2019.

By 2020, the Reykjavík Capital Area comprised two-thirds of the Icelandic population. This has involved a decisive shift in political, economic and cultural hegemony as nearly all the larger Icelandic businesses, political leaders, governmental institutions, cultural institutions and creative class, media outlets and institutions of higher education have concentrated in the 1% of the landmass covered by the capital area. About half of the remaining non-metropolitan population lives in either the exurban regions within 100 km of the city or the northern regional centre of Akureyri. In addition to the growth of the Reykjavík Capital Area, substantial urbanisation has taken place in other regions. Farming communities and small fishing villages represent about 10% of the national population, but these communities are the site of most films set outside the Reykjavík Capital Area.

The volume of film production in Iceland is quite remarkable given that the 2020 national population of 364,000, and the Reykjavík Capital Area population of 233,000 is comparable to the metropolitan areas of, for example, York in England, Örebro in Sweden, Cork in Ireland or South Bend, Indiana. In this small population, the market share of Icelandic movies is only about 5% against the 91% US market share (Icelandic Film Centre 2020b). The annual revenue from domestic ticket sales for Icelandic films in the period 2014–2019 thus only ranged from ISK 54–164 million (EUR 0.4–1.0 million). The total annual revenue of the Icelandic film industry, however, is about ISK 13,000 million (EUR 80.9 million).

The key to this apparent paradox lies in the fact that the Icelandic film industry successfully caters to the international niche market for independent and artistic film, in part with the support of the European Cinema Support Fund (see, e.g., Schindler 2014; Norðfjörð 2019). While this is only a small part of the global film industry, it is an enormous market for the small local industry operating in the tiny Icelandic context. In the period 2013–2019, Icelandic films won between 31 to 103 awards annually at international film festivals (Icelandic Film Centre, 2020a). Interestingly, the vast majority of these critically acclaimed films were set in rural Iceland.

Icelandic urban and rural films

According the Icelandic Film Centre's database of all Icelandic movies (Kvikmyndavefurinn 2020), a total of 214 predominantly Icelandic feature films were released in the period 1918–2019. As shown in Figure 10.1, more than half of all Icelandic films continue to be set in rural Iceland or have a strong rural dimension. While the average number of Icelandic films released increased over time, the proportion of films set in rural Iceland remained similar. Of the 78 Icelandic feature films that premiered in 2010–2019, 42% were predominantly set in rural areas, compared to 47% of the 30 films premiered in 1980–1989.

The rurality of contemporary Icelandic films tends to have a somewhat disorienting, asynchronic quality of being apparently set in modern times, yet being reliant on radio news, landline telephones, 1940s Ferguson tractors and strikingly similar, poison-green kitchens reminiscent of the 1960s. Immigrants are largely absent from rural communities and the female population is mostly limited to one strong grandmother and/or one object of love for the protagonist. The Icelandic rural, it seems, is caught in a perpetual time warp, unable to either perish with its bygone era or adjust to the modern

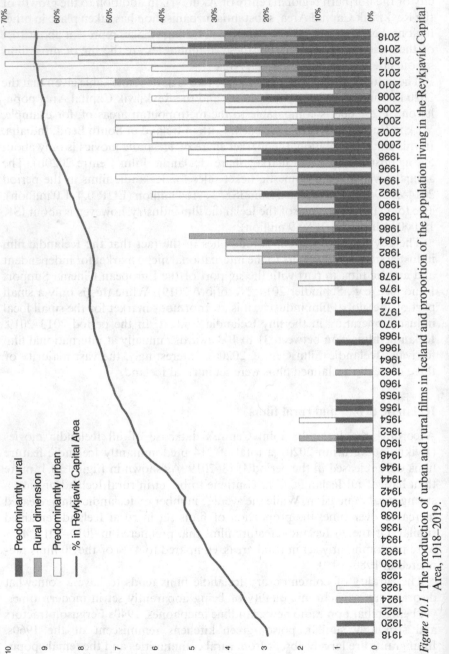

Figure 10.1 The production of urban and rural films in Iceland and proportion of the population living in the Reykjavik Capital Area, 1918–2019.

times. This resembles the supposedly timeless qualities of the essentially English rural idyll – frozen in time – but with a dark and dystopic twist.

A surprising share of Icelandic films are set the in the Westfjords of Iceland, a region of deep fjords, crumbling mountains and small, isolated fishing communities. The Westfjords are also the only region of the country to have suffered substantial population loss over the past century. In 1918, there were over 13,000 people in the Westfjords, but by 2020 the population had dwindled to about 7,000 or 5% of the non-metropolitan population of Iceland. About 19% of the inhabitants are immigrants, well above the Reykjavík Capital Area average of 15% (Statistics Iceland 2020b). Films set in this dramatic environment include two historical period dramas,[1] two family films,[2] three horror movies,[3] four comedies[4] and six contemporary dramas.[5] Remarkably, the Westfjords of Icelandic cinema is exclusively populated by ethnic Icelanders. In contrast, Icelandic film are hardly ever set in the more diverse, growing regions or larger towns that account for the majority of the non-metropolitan population.

Film production being a predominantly urban industry, it is hardly surprising that the city of Reykjavík has been a recurring setting for Icelandic films. While the city frequently serves as a passive background to stories about the joys and sorrows of modern life, the supposedly brutal underworld and mean streets of Reykjavík has been portrayed in a long line of Icelandic action thrillers.[6] This theme of Reykjavík as an urban jungle of hard drugs, organised crime and violence is all the more remarkable given the relatively peaceful nature of the small, peripheral city in global terms. The annual murder rate of 0.7 per 100,000 in Reykjavík and 0.6 for Iceland as a whole is far below the 2.7 average across European countries (UNODC 2020). In fact, the only European countries with a lower murder rate than Iceland are Andorra, San Marino and the Holy See. Similarly, only five European countries rank lower than Iceland in terms of lifetime prevalence of illicit drug use among youth (ESPAD 2015).

City-based comedies include such diverse themes as a young hipster in love with his mother's lesbian lover (*101 Reykjavík* [2000]), girls navigating Reykjavík nightlife (*Dís* [2004]), the day everything went wrong in the life of a middle-aged teacher (*Jóhannes* [2009]), a dead, life-thirsty grandfather haunting his old house (*Ófeigur gengur aftur (Spooks and Spirits)* [2013]) and an out-of-control suburban conflict over a shadow-casting tree (*Undir Trénu (Under the Tree)* [2017]). In contrast, rural comedies almost exclusively involve city people out of their comfort zone, such as a pop band on a tour of the countryside (*Með allt á hreinu (On Top)* [1982]), a busload of wedding guests looking for a white church with a red roof in a countryside full of churches (*Sveitabrúðkaup (Country Wedding)* [2008]), young men taking temporary jobs in the extraction industries, such as fish processing (*Nýtt líf (New Life)* [1983]) or farming (*Dalalíf (Valley Life)* [1984]) and an hapless urbanite displaced in rural Iceland as a primary school teacher (*París norðursins (Paris of the North)* [2014]), golf course attendant (*Albatross* [2015]) or manager of a failed slaughterhouse (*Kurteist fólk (Polite People)* [2011]).

More dramatic, critically acclaimed and award-winning Icelandic films, however, tend to draw a darker picture of rural Iceland and its inhabitants, including rampant alcoholism, emotional repression, sexual abuse, quiet desperation, claustrophobia and hostile and unforgiving nature. It is important to note that a "rural idyll" has never been a prominent feature of Icelandic film, and while Icelandic nature has been shamelessly promoted from the outset, rural communities have almost invariably been a better place to *be from* than to *be in*. In more recent times, however, the bleakness of rural communities thrown in sharp relief against majestic and imposing mountains has become something of an Icelandic film genre in international cinema.

In critically acclaimed Icelandic films, toxic masculinity emerges as the fundamental problem of rural communities, both for rural men having trouble with themselves and for rural women having trouble with their men. Films such as Dagur Kári's (2003) *Noi Albinoi* (*Noah the Albino*),[7] Rúnar Rúnarsson's (2015) *Þrestir* (*Sparrows*)[8] and Guðmundur Arnar Guðmundsson's (2016) *Hjartasteinn* (*Heartstone*)[9] portray the existential angst of young boys growing up in dysfunctional rural families and dysfunctional rural communities. The fathers are invariably emotionally stunted and depressed alcoholics and the mothers are either emotionally absent or have literally left the rural community. Rural adolescent societies are alternatively portrayed as almost nonexistent (*Nói Albínói*), dominated by alcohol, violence and sexually abusive adults (*Þrestir*), or unsupervised hotbeds of bullying, slut-shaming and homophobia (*Hjartasteinn*).

Beyond the supporting roles of the blighted fathers of young men in crisis, critically acclaimed cinematic portrayals of ageing rural men take various forms. The old farmer in Friðriksson's (1991) *Börn Náttúrunnar* (*Children of Nature*)[10] burns his photo albums, shoots the dog and plays his organ one last time before moving to a retirement home in Reykjavík. In Grímur Hákonarson's (2015) *Hrútar* (*Rams*),[11] two ageing brothers who live on adjacent farms have not spoken in decades due to some undefined and perhaps long-forgotten reason. More flamboyantly, the rural men of Benedikt Erlingson's (2013) *Hross í Oss* (*Of Horses and Men*)[12] perish one by one from alcohol poisoning, tractor road-rage and other dumb ways to die, cut their eyes on barbed wire, and abuse and kill horses for their own social and physical well-being.

Young rural women in Icelandic films tend to be plagued by the locker-room mentality of local boys (*Hjartasteinn*), the lack of promising boyfriend material (*Nói Albínói*) and adult sexual predators (*Þrestir*), while middle-aged rural women tend to have given up on hopeless men and hopeless communities and left for the city. In *Hross í Oss*, however, the middle-aged rural women use binoculars to quietly observe the vanity, alcoholism, violence and plain stupidity of their rural husbands, and the growing number of widows make googly eyes at the only eligible middle-aged bachelor in the community. Elderly rural women tend to be strong-willed paternal grandmothers who provide both physical and emotional shelter for their sons and grandsons and attempt in vain to keep fathers and sons together when they have been abandoned by their wives and mothers (*Nói Albínói* and *Þrestir*).

In these critically acclaimed movies, the unbearable sadness of dying communities generally culminates in either doom or out-migration. In *Börn Náttúrunnar*, the deeply unhappy retired farmer reunites with an old flame in the retirement home and together they flee the "dump in the South" in a stolen vehicle. Repeatedly dodging the police, the old couple make it all the way back home to die. In *Hrútar*, the brothers who have not spoken in decades are finally reunited when a female, urban, university-educated vet determines that the entire family sheepstock must be slaughtered to eliminate a lethal, highly contagious disease. After personally killing most of the flock and failing to hide the few remaining sheep in the basement, the brothers flee to the highlands where they perish naked in a snowstorm.

Even when the protagonists survive, they generally do so at enormous personal cost. After several hilariously failed attempts to leave the village, *Nói Albínói* retreats to the shelter of his grandmother's basement, thereby narrowly escaping death in the avalanche that wipes out most of the village. In *Hjartasteinn*, a young boy is physically abused by his alcoholic father, bullied and abused by his peers, and confused by his homosexual attraction to his best friend. After attempting but failing to kill himself with a shotgun, moving to Reykjavík becomes the only viable escape. In *Þrestir*, a young man is sexually abused at home by an older woman and proves unable to prevent the rape of his unconscious girlfriend by adult partygoers. He eventually gives up on his dream of returning to Reykjavík and in the final scene settles tenderly into the arms of his half-dressed, passed-out father on the sofa of his deceased grandmother's house.

Internal and external orientalism

Between 1918 and 2020, the non-metropolitan population of Iceland grew from 77,000 to 131,000 in absolute terms, but declined in relative terms from 80% of the national population in 1918 to 34% in 2020. This involved a massive transfer of political, economic and cultural power from the provinces to the Reykjavík Capital Area. This socio-economic and demographic transformation has spurred persistent misconceptions of a massive decline in absolute numbers (Bjarnason 2020) and repeated claims that Iceland has now become a "City State" (*Borgríki*) with a negligible and irrelevant rural population (see, e.g., Sveinsson 2001; Einarsson 2008).

The perceived gap between the densely populated city of the future and the sparsely populated countryside of the past has grown as first-generation migrants from rural areas have become a smaller proportion of the urban dwellers. This has been accompanied by a cultural transformation and dramatic shift in the imagery of the countryside in popular culture, as exemplified by the movies produced by the Icelandic film industry. The small city of Reykjavík has thus been framed in terms of the excitement, diversity and danger of an international metropolis, and as the only hope for Iceland to retain young people and succeed in an urban, competitive global economy. In contrast, the rural is exemplified by lonely, ageing bachelor farmers and tiny, isolated fishing villages, rather than organic regions of interrelated large and small towns, subdivisions, villages, hamlets, farmsteads and single houses. It is a rural where women are scarce and immigrants are non-existent.

Furthermore, the problems of the rural do not appear to be national-level disinvestment in business and public services, poor roads, imploding social services, expensive heating, expensive transport, unreliable electricity, lack of access to health care, national fisheries regulations or any of the other issues weighing heavily on rural residents. The rural is instead plagued by toxic masculinity, alcoholism, violence, disintegrated communities, harsh nature and ultimately the lack of personal drive to leave for the city. Thus, existing everyday problems are downplayed in the same way that the English rural idyll masks poverty and social exclusion in England (Bell 2006; Milbourne 2014) – with the critical difference that in Icelandic cinema the dominant narrative resisting social change is framed by dystopia rather than utopia.

A number of film scholars have remarked upon the large proportion of Icelandic films set in rural communities (e.g. Crowie 2005; Gravestock 2019; Konefał 2014; Møller 2005; Vilhjálmsson 2019). However, this has generally been attributed to nationalism, recent urbanisation, the rural roots of many film-makers, the need to demolish a pervasive myth of the rural idyll, or simply the cinematic mountains of the Westfjords. However, even as rural Iceland has faded ever further into the background of the national discourse and media coverage, the fascination of Icelandic film-makers with rural Iceland has not diminished. The steady stream of films about broken, desperate people in the smallest and most vulnerably rural communities may thus have deeper roots in contemporary Icelandic society in general and among the Reykjavík creative classes in particular.

Jansson (1995) has argued that films such as Alan Parker's [1988] *Mississippi Burning* serve as a tool of internal orientalism to construct and reinforce images of the US South as a region of racism, violence and poverty as the "other", in contrast to a dominant American national identity of tolerance, justice and middle-class comforts. In a parallel fashion, Eriksson (2010) argues that Kjell Sundvall's (1996) *Jägerna* (*The Hunters*) draws a strong distinction between the backward, intolerant, violent and welfare-dependent residents of Northern Sweden (Norrland) and the progressive, tolerant, peaceful and productive urban Stockholmers. In the case of the Icelandic film industry, however, these contrasts are only partially being drawn for the benefit of an Icelandic audience. The international niche market for independent film is crucial for the impressive volume of Icelandic film being produced each year, even though some films are primarily produced for the domestic market. The perceived authenticity of the rural communities portrayed in Icelandic film has thus been commodified and successfully marketed to audiences and film critics abroad, and this authenticity draws on notions of Icelandic 'otherness' and well-rehearsed narratives about Icelandic culture and life being remote, marginal and exotic (see, e.g., Ellenberger 2007). Icelandic film can thus simultaneously be seen as an instance of internal orientalism within Iceland and external orientalism in the global context.

Beyond the simple economic imperative, the wildly exaggerated difference between the cool, rough and gang-infested Reykjavík and the quiet desperation of rural Iceland may serve to position the film-makers themselves comfortably *vis-à-vis* their peers abroad. Iceland itself is of course geographically peripheral in Europe, and the national population accounts for only 0.05% of the population of the continent (United Nations 2020). In a sense, Iceland is thus the Westfjords of Europe and, in the process of globalisation and European integration, Icelandic film-makers may have turned to internal orientalism in order to position themselves at the core of European cinema by exaggerating the exotic, supposedly authentic Icelandic rural from the vantage point of the cool, cosmopolitan city of Reykjavík.

Conclusion

While the "rural idyll" has never been a hallmark of Icelandic cinema, representations of rural people have become increasingly dark, sinister and unruly. In this chapter we have reviewed the evolution of the film industry and its relationship to urbanisation, unpacking the dominant narratives about rural society and exploring what this Icelandic commodification of rural authenticity might suggest for policy and politics.

It should be emphasised that film-makers and other storytellers are of course under no obligation whatsoever to provide, collectively, a fair and balanced representation of society. The constant reiteration of cinematic stereotypes is nevertheless a powerful form of the social construction of reality. The world of Hollywood movies thus tends to be overpopulated by cops, lawyers, beautiful young people in love, criminal masterminds and their vicious henchmen, and underpopulated by electricians, accountants, siblings without rivalry, as well as the old, the disabled and the obese. In the same fashion, the rural communities of critically acclaimed Icelandic film are overpopulated by severely alcoholic fathers and deeply unhappy bearded men, and underpopulated by young women, immigrants and happy families. Furthermore, the broad range of Icelandic settlements, including downtown Reykjavík, residential neighbourhoods, exurban regions and more rural towns, villages, hamlets and farming communities, has essentially been reduced to a razor-sharp distinction between a modern, diverse, vibrant and occasionally dangerous global city and a desperate, claustrophobic, lonely and pathetic but visually impressive remote rural.

The images of rural Iceland carefully crafted by urban Icelandic film-makers simplify and inflate urban–rural differences and throw a sophisticated and cosmopolitan image of urban Reykjavík into stark relief. In this sense, the films serve as instruments of internal orientalism that reflects and reinforces the cultural, political and economic hegemony of the city over the provinces by perpetuating the notion of the manifest destiny of Iceland as a future city-state and recasting structural inequalities as the consequences of personal shortcomings and poor residential choices. Interestingly, this takes place through the commodification and international marketing of rural dystopia as an authentic, unsparing view of the harshness of Icelandic nature and rural culture. This external orientalism that capitalises on international stereotypes of Icelanders as rough, heavy-drinking, bearded men barely surviving in the harshness of majestic nature has brought significant professional rewards to the predominantly bearded Icelandic film-makers.

Notes

1 *Útlaginn* (*Outlaw*; 1981) and *Myrkrahöfðinginn* (*Prince of Darkness*; 1999).
2 *Ikingut* (2000) and *Duggholufólkið* (*People of Dugghola*; 2007).
3 *Blóðrautt sólarlag* (*Blood-red Sunset*; 1977), *Skammdegi* (*Midwinter*; 1985) and *Ég man þig* (*I Remember You*; 2017).
4 *Albatross* (2015), *Heiðin* (*The Heath*; 2008), *París Norðursins* (*Paris of the North*; 2014) and *Á annan veg* (*Either Way*; 2011).
5 *Börn náttúrunnar* (*Children of Nature*; 1991), *Ingaló* (1992), *Í faðmi hafsins* (*In the Arms of the Sea*; 2001), *Nói Albínói* (2003), *Veðramót* (*Quiet Storm*; 2007) and *Þrestir* (*Sparrows*; 2015).
6 For example, *Blossi* (*Flare*; 1997), *Sporlaust* (*No Trace*; 1998), *Mýrin* (*Jar City*; 2006) (*Reykjavík-Rotterdam*; 2008), *Borgríki* (*City State*; 2011), *Svartur á Leik* (*Black's Game*; 2013), *Blóð Hraustra Manna* (*Brave Men's Blood*; 2014), *Austur* (*East*; 2015) and *Vargur* (*Wolf*; 2018).
7 24 nominations at international film festivals, including 15 awards.
8 27 nominations at international film festivals, including 10 awards.
9 43 nominations at international film festivals, including 26 awards.
10 Nominated for an Oscar award as the best Foreign Language Film in 2002.
11 A total of 34 nominations at international film festivals, including 22 awards.
12 A total of 23 nominations at international film festivals, including 15 awards.

References

Alessio, D. and Jóhannsdóttir, A. L. (2011). Geysir and "girls": Gender, power and colonialism in Icelandic tourist imagery. *European Journal of Women's Studies*, 18, 35–50.
Alþýðublaðið, 1949. "Milli fjalls og fjöru" fær góðar viðtökur í Danmörku. *Alþýðublaðið*, May 25th, 1949, p. 7.
Bartlett, F. 1923. *Psychology and primitive culture*. Cambridge: Cambridge University Press.
Bell, D. 2006. Variations on the rural idyll, in Cloke,P., Marsden,T., and Mooney,P. H. (eds), *Handbook of rural studies*. London: Sage, pp. 149–160.
Berger, P. L. and Luckmann, T. 1966. *The social construction of reality: A treatise in the sociology of knowledge*. Garden City, NY: Anchor Books.
Bjarnason, T. 2020. The myth of the immobile rural: The case of rural villages in Iceland. In Lundmark, L., Carson, D., Eimmerman, M. (eds.). *Dipping into the north: Living, working and traveling in sparsely populated areas*. Houndmills: Palgrave Macmillan, pp. 55–70.
Brandth, B. and Haugen, M. S. 2011. Farm diversification into tourism – implications for social identity? *Journal of Rural Studies*, 27(1), 35–44.
Crowie, Peter. 2005. *Icelandic films*. Reykjavik: The Icelandic Film Centre.
Dagur, 1962. Fyrsta íslenska kvikmyndin. *Dagur*, 20th October 1962, p. 7.

Ellenberger, Í. 2007. *Íslandskvikmyndir 1916–1966: Ímyndir, Sjálfsmynd og Vald.* Reykjavik: Sagnfræðistofnun HÍ.

Eriksson, M. 2010. "People in Stockholm are smarter than countryside folks" – reproducing urban and rural imaginaries in film and life. *Journal of Rural Studies,* 26, 95–104. doi:10.1016/j.jrurstud.2009.09.005.

ESPAD. 2015. *The 2015 ESPAD report: Results from the European school survey project on alcohol and other drugs.* Council of Europe, accessed September 1, 2020 at http://espad.org/report/home/

Fontaine, A. S. 2006. "Quentin Tarantino, you should work for the Icelandic tourist board!" *Reykjavík Grapevine,* 13 January 2006. Accessed August 2, 2020 at https://grapevine.is/mag/articles/2006/01/13/quentin-tarantino-you-should-work-for-the-icelandic-tourist-board/

Forslund, B. 1988. *Victor Sjöström: His life and his work.* New York: Zoetrope.

Foucault, M. 1970. *The order of things: An archaeology of the human sciences.* New York: Pantheon Books.

Foucault, M. 1977. *Discipline and punish: The birth of the prison.* New York: Pantheon Books.

Foucault, M. 1978. *The history of sexuality volume 1: An introduction.* New York: Pantheon Books.

Fram, 1920. Samkomuhús. *Fram,* 17 January, p. 10.

Gramsci, A. 1971. *Selections from the prison notebooks of Antonio Gramsci,* edited and translated by Hoare, Q. and Smith, G. N. London: Lawrence & Wishart.

Gravestock, S. 2019. *A history of icelandic film.* Waterloo, Ontario: Wilfrid Laurier University Press.

Hechter, M. 1975. *Internal colonialism: The celtic fringe in British national development, 1536–1966.* Berkeley: University of California Press.

Huijbens, E. H. 2011. Nation branding – a critical evaluation, in Ísleifsson, S. (ed.), *Iceland and images of the north.* Québec: Presses de l'Université du Québec, pp. 553–582.

Icelandic Film Centre. 2020a. International awards: Awards for Icelandic films at international film festivals and award ceremonies. Accessed August 2, 2020 at www.icelandicfilmcentre.is/facts-and-figures/international-awards/

Icelandic Film Centre. 2020b. *Key Figures.* Accessed July 12, 2020 at www.icelandicfilmcentre.is/facts-and-figures/key-figures/

Jansson, D. R. 1995. "A geography of racism": Internal orientalism and the construction of American national identity in the film Mississippi Burning. *National Identities,* 7, 265–285.

Konefał, J. S. 2014. City, countryside and nature as discursive devices used to strengthen the national identity in Icelandic cinematography. *Studia Humanistyczne Agh,* 13(3), 61–79.

Kvikmyndavefurinn. 2020. *Icelandic Feature Films.* www.icelandicfilms.info/films/bytype/genre/movie

Landið, 1918. Fjalla-Eyvind. *Landið,* 14 June 1918, p. 94.

Marcuse, H. 1964. *One-dimensional man: Studies in the ideology of advanced industrial society.* Boston: Beacon Press.

Milbourne, P. 2014. Poverty, place, and rurality: Material and sociocultural disconnections. *Environment and Planning A,* 46, 566–580.

Morgunblaðið, 1980. Land og synir. *Morgunblaðið,* 27 January, p. 15.

Møller, B. T. 2005. In and out of Reykjavik: Framing Iceland in the global daze, cinema and nation, in Nestingen, A. and Elkington, T. G., eds., *Transnational cinema in a global north: Nordic cinema in transition.* Detroit: Wayne State University Press, pp. 307–340.

Nerlich, B. and Döring, M. 2005. Poetic justice? Rural policy clashes with rural poetry in the 2001 outbreak of foot and mouth disease in the UK. *Journal of Rural Studies,* 21(2), 165–180.

Norðfjörð, B. Æ. 2019. Ljós í myrkri: Saga kvikmyndunar á Íslandi. *Ritið,* 19(2), 19–42.

Piaget, J. 1932. *The moral judgment of the child.* London: Routledge & Kegan Paul.

Pratt, A. C. 2007. "Imagination can be a damned curse in this country": Material geographies of filmmaking and the rural, in Fish, R (ed.), *Cinematic countrysides.* Manchester: Manchester University Press, pp. 127–146.

Schindler, A. 2014. State-funded Icelandic film: National and/or transnational cinema? in Jones, H. D. (ed.), *The media in Europe's small nations.* Cambridge: Cambridge Scholars Publishing, pp. 69–85.

Statistics Iceland. 2020a. *Sögulegar hagtölur: Mannfjöldi.* Accessed 7 August 2020 at https://sogulegar.hagstofa.is/mannfjoldi/

Statistics Iceland. 2020b. *Innflytjendur eftir kyni og sveitarfélagi 1. janúar 1996–2019.* Accessed 7 August 2020 at https://hagstofa.is/talnaefni/ibuar/mannfjoldi/bakgrunnur/

Sutherland, J.-A. and Kathryn M. Feltey (eds.). 2013. *Cinematic sociology: Social life in film.* Thousand Oaks, CA.

Sveinsson, J. R. 2001. Landsbyggðin - jaðarsvæði borgríkisins? *Morgunblaðið,* 23 January 2001, p. C-30.

UNODC. 2020. *Homicide.* United Nations Office Drugs and Crime. Accessed 6 August 2020 at https://dataunodc.un.org

United Nations. 2020. *Population Data.* Accessed 31 August 2020 at https://population.un.org/wpp/Download/Standard/Population/

Útvarpstíðindi. 1949. Milli fjalls og fjöru. *Útvarpstíðindi,* 7 March, 1949, p. 67.

Vilhjálmsson, B. T. 2019. Frá sveitabænum að stafrænu byltingunni. *Ritið,* 19(2), 1–18.

Weber, M. 1946. Politics as a vocation, in *From Max Weber: Essays in sociology,* translated and edited by Gerth H. H., and Mills, C. W. New York: Free Press, pp. 77–128.

11 Rural authenticity as cosmopolitan modernity?

Local political narratives on immigration and integration in rural Norway

Guro Korsnes Kristensen and Berit Gullikstad

Introduction

What happens to notions of rurality when immigrants from all over the world settle and make homes for themselves in rural communities? How do local majority understandings concerning immigration and immigrant integration in small municipalities contribute to the production of rural authenticity? In this chapter, we explore the ways in which the mayors of 13 relatively small municipalities in a rural region of Norway narrate their experiences with immigration and integration. As elected political representatives, mayors are positioned at the crossroads of rural and urban imperatives: they can be expected to defend and seek support for *local* political choices, such as settling refugees and supporting industries in need of labour, while also experiencing *inter-municipality* and *national* pressures to merge into larger entities. With this in mind, it is reasonable to read their narratives about immigration and immigrant integration as legitimising of local political choices within a broader context of a neoliberal policy in which size matters, casting "small" rural areas in a negative light and pushing smaller municipalities to merge with others. Thus, the mayors' stories can be read in a rural–urban framing that makes salient the tensions involved in the different levels of decision-making.

In the last 20 years, immigration and integration in rural communities have gained increasing attention (Woods 2012; McAreavey and Argent 2018; Rye and Scott 2018). In this context, particular interest is paid to rural regions and communities with little prior experience of migrations: so-called New Immigration Destinations (McAreavey and Argent 2018). In the Northern European context, this shift followed in the wake of the expansion of the EU in 2004 and 2007 and the refugee crisis of 2015 onwards which, in addition to lifestyle migration, have led to a population increase in many rural areas (Søholt, Tronstad and Bjørnsen 2014; Åberg and Högman 2015; Rye 2018; Karlsdóttir et al. 2020).

How this situation has affected rural places and communities and if/how work migrants, refugees and their families have been welcomed as part of the community, have become significant questions in both rural studies and migrant integration studies through the so-called local turn (see, e.g.,

DOI: 10.4324/9781003091714-13

McAreavey and Argent 2018; Meissner and Heil 2020), whereby local processes and policies are regarded as more inclusive than those at the national
level (McAreavey and Argent 2018; Hadj Aboud 2019). While some research
papers report that smaller municipalities are welcoming new migrants
because they lead to increasing populations and increased wealth, and hence
contribute to the maintenance or even regeneration of local communities
(Steen 2009; Arora-Jonsson 2017; Aure et al. 2018; Rye 2018; Søholt et al.
2018; Steen and Røed 2018), other papers have stressed that integration is
particularly challenging in rural societies because of a lack of social diversity,
the existence of racism and fewer possibilities for education and work (Ray
and Reed 2005; Valenta 2007; Arora-Jonsson 2017; Rye and Scott 2018).

The same opposition can also be found in the Norwegian media. There
have been several debates around whether rural areas are better or worse than
urban areas when it comes to integration (Amundsen 2017; Nicolajsen 2019;
Svensson 2019), and some local communities have received attention for their
success in including newcomers into the community and/or their willingness
to receive more. A telling example of the latter is a news item published by the
Norwegian public broadcaster NRK in 2017. When the number of refugees
to Norway started to drop, the news described three municipalities as being
engaged in a "battle for refugees" (Befring 2017). In this report, the Norwegian
public was introduced to three very different municipalities that all wanted to
welcome more refugees. The most rural one presented an idyllic picture of a
harmonious and warm community where the refugees who had settled there
were already feeling at home.

Both the research and the media reports touch upon the long and continuing rural–urban debates in which notions of rurality and rural identity as
"rural idyll", "rural resilience", "white rurality" or "rural globalisation and
cosmopolitanism" are in play (Berg and Lysgård 2004; Hedberg and
Haandrikman 2014; Søholt et al. 2018; Woods 2018). Rural identity or
authenticity has become important in making the "new local community"
(Almås et al. 2008) into an attractive option (Woods 2011; Lysgård and
Cruickshank 2016) in terms of employment opportunities, welfare services,
voluntary work, cultural events, gender equality and globalised relations
(Almås et al. 2008).

In this chapter we explore how leading local politicians – who, on the one
hand, are dealing with national policies and, on the other, with the specific
local context – reflect upon their municipality's experiences with immigration
and integration. Which narratives about the local community, "the locals"
and rurality do these reflections produce? And what conceptions of rural
authenticity are embedded in these narratives?

Immigration, integration and rurality

According to Rye, the classical theme in rural migration studies has been
rural-to-urban migration within the framework of the nation-state (Rye
2014). In recent years, however, scholarly interest has shifted towards rural

in-migration, back-to-the-land movements, rural gentrification and counter-urbanisation (Jentsch 2007; Rye 2014, 2018). There is also a growing body of literature focusing specifically on labour immigration (Aure et al. 2018; Rye 2018; Rye and Scott 2018) and settlement of refugees and asylum-seekers in rural areas in Europe (Hubbard 2005; Finney and Robinson 2009; Bygnes 2017; Søholt and Aasland 2019).

A recurring theme and persistent question in this research is whether rural societies are less tolerant of in-migration than their urban counterparts. How rural societies deal with this new in-migration and integration can be explored in the contested terms of rurality, identity and values, which are usually constructed on the rural–urban dimension (Haugen and Lysgård 2006). In this respect, Oliva's term "rural melting pot" (Oliva 2010) is significant, as it draws attention to the ways in which international migration challenges and transforms traditional rural values and ways of life. An example of this is presented in Hedberg and Haandrikman's investigation of rural societies in Sweden, in which they discuss how the influx of international labour migrants challenges what they label as "white rurality" in racial terms and the "rural idyll" in class-based terms (2014). In line with this, "nativeness", rootedness and authenticity are described as qualities aligned with the Northern European countryside (Søholt et al. 2018, p. 21), as is a particularly strong interconnectedness between identity and place/locality in rural areas (Åberg and Högman 2015).

As rural areas are often seen as more homogeneous and less dynamic than urban areas (Haugen and Villa 2006; Milbourne and Kitchen 2014), immigration could present a challenge. Several researchers have argued that immigrants will fit in more easily and feel a greater sense of belonging in an urban environment than a rural environment, as rural areas and small places where "everyone knows everyone" can generate a higher degree of social exclusion, xenophobia and suspicion towards foreigners (Neal 2002; Ray and Reed 2005; Valenta 2007). Later studies focusing on labour immigrants in rural areas also stress the clear tendencies of social fragmentation, polarisation, contradictions and little contact between migrants and "old" inhabitants (Arora-Jonsson 2017; Aure et al. 2018; Rye 2018; Rye and Scott 2018; Villa 2019).

On the opposite side, the rural is also seen as a site for openness, mobilities and cosmopolitanism (Woods 2018; Villa 2019). For example, Kasimis et al. (2010)) claim that less-developed rural regions are generally very willing to accept immigrants into the local community. Even though rural places are typically associated with stability (Milbourne and Kitchen 2014), recent research has argued that mobility is and has always been central to the enactment of the rural (Milbourne and Kitchen 2014; Aure et al. 2018). Moreover, according to recent Norwegian research, heterogeneity is important for local communities, as it makes them more attractive to the desired workforce, as well as promoting population growth (Lysgård and Cruickshank 2016; Aure et al. 2018; Rye 2018). Similarly, where labour migration counters trends toward rural depopulation through a so-called "demographic refill" (Hedberg and Haandrikman 2014, p. 129) and economic growth (Aure et al. 2018; Rye 2018;

Steen and Røed 2018), there is an assumption that incomers will be more easily accepted.

Traditionally, these two approaches to the rural society have been seen as mutually exclusive. Søholt et al. (2018) argue instead that both can be seen as part of a wider scale wherein variations are differently emphasised and paralleled in order to maintain local resilience. In line with this, Kristensen and Sætermo (2021) argue that integration processes are place-specific in ways that potentially transgress the rural–urban divide.

In the Norwegian context, the situation of depopulation is highly relevant (Aasbrenn and Sørlie 2016). At the same time, Norwegian regional policy has had an overall aim of ensuring equal living conditions for people in all parts of the country and of maintaining the existing population pattern (Haugen and Lysgård 2006, p. 174). In line with this, "rurality" is claimed to have had high prestige in Norway, and rural areas have far more political power than might be expected, given the size of the rural population (Haugen and Lysgård 2006, p. 174). This regional policy is also visible in the policies regulating the settlement of refugees and asylum-seekers, where dispersed settlement has been used to contribute to preserving the main settlement pattern since 1990 (Søholt et al. 2018; Tønnesen and Andersen 2019). Today, immigrants are represented in all Norwegian municipalities, and even small communities that have for years suffered from depopulation are experiencing population growth as well as economic growth due to immigration (Aure et al. 2018; Berg-Nordlie 2018; Rye 2018; Søholt et al. 2018). In some municipalities, labour migrants account for a substantial part of the population, while in others there are more refugees. Whereas labour immigration is a consequence of the labour market and the local industry in a region, the settlement of refugees is a political decision by the local government which can decide if, and how many, refugees can be settled. The increase in the number of people being settled during 2015 and 2016 placed strains on the capacity of the municipal services, and many municipalities were reluctant to settle refugees. Thus, the central government provided economic incentives for the five years following settlement, causing many municipalities – including small municipalities in rural areas that had little or no previous experience with refugee settlement – to welcome rather high numbers of refugees (Søholt and Aasland 2019; Tønnesen and Andersen 2019). When the number of refugees coming to Norway decreased in 2017 and 2018 due to more restrictive national policies, the situation turned to the aforementioned "battle for refugees".

According to researchers, such as Steen and Røed (2018) and Søholt et al. (2018), the positive attitude towards immigration among local politicians and municipality bureaucrats in rural areas is to a great extent a consequence of self-interest, since immigration contributes to development and growth. Any emotional worries about cultural exchange are largely repressed (Steen 2009; Søholt et al. 2018). Moreover, it is expected that the immigrants will "hav[e] the right attitude, tak[e] initiative, [join] local organisations, [develop] language skills, and mak[e] an effort" (Søholt et al. 2018, p. 226). Hence, responsibility for integration is primarily put on the immigrants, and, from the rural elite's point of view, to a great extent successful integration means that the immigrants are becoming more like the majority (ibid., p. 229; Gullikstad and Kristensen 2021). Another part of the picture is that refugees and asylumseekers, despite economic incentives, are often described as financial burdens on the municipalities, as well as potential security risks for the local communities (Berg-Nordlie 2018; Søholt et al. 2018).

Methods and analytical tools

The research was conducted in a region that can be described as predominantly rural, in the sense that a majority of the territory's 38 municipalities are sparsely populated, with mostly small towns and villages. In line with the country as a whole, this region is also clearly affected by increased international immigration. The proportion of immigrants, which includes labour migrants, refugees and lifestyle settlers, is 11% for the whole region, varying from 7–24% between different municipalities (IMDi 2020). In the wake of the so-called refugee crisis of 2015, quite a few smaller municipalities settled refugees for the first time.

In this region, we conducted qualitative interviews with 13 mayors in relatively small municipalities (numbers of inhabitants ranging from 2,600 to 15,000). The interviews took place in 2018 and 2019, as part of a qualitative research project on immigrant integration in contemporary Norway.[1] The selected municipalities are both coastal and inland, and they have a wide range of business activities, with fishing and farming as dominating industries. The political leadership of the municipalities where we did our research was mostly from the labour party and the agrarian party.

In Norway, the mayor is the highest-ranking official in a municipality government. This means that the mayor is a leading figure in the local politics of the municipality. Moreover, mayors often have a long history as local politicians before entering this position. As such, they tend to be well informed about the municipality, including its history and practical and ideological rationales for political decisions, as well as ongoing political debates along a wide variety of themes, including the actual and potential effects immigration might have on the specific municipality.

The aim of the interviews was to encourage the mayors to speak as freely as possible about the municipality's experiences with immigration and integration. Overall, the mayors were active in the interviews, and the strategy of conducting open interviews – wherein the interviewees were free to define the direction in which the conversation would go – worked very well.[2]

When analysing the interview transcripts, our focus was on cultural meaning-making among policymakers – not policy-making as such. This means that our aim was not to identify the correct chronology, genuine motivation, true effects, etc. Rather, we wanted to explore how particular situations and (hi)stories are made meaningful and culturally acceptable in the particular context of the research interview, which we expect to also reflect cultural meaning-making taking place in other public and political settings. We argue that an important way of capturing cultural meaning-making is to focus on the stories that are (re)produced when people meet and interact in different everyday settings and contexts. In this respect, we draw on narrative analysis (Riessman 2008). An important feature of narrative analysis is that the production of stories is seen as a social activity, in the sense that the "smaller" stories people produce individually are also part of the "bigger" stories that societies and cultures share (Bo et al. 2016). In this chapter, we refer to these socially produced and shared stories as "collective narratives". While individuals are not in thrall to the collective narratives available to them, they are influenced by them, and draw on them in their everyday sense-making activities, including their own storytelling. Hence, narrative research provides information about individual meaning-making and the wider social context in which this meaning-making takes place.

According to Lotta Junnilainen, narratives are historically informed collective processes of place-making that, once dominant in public discourse, affect what defines "the community" and what does not (2020). These place-making narratives can be seen as influential meso-level narratives that are providing models of the place, who belongs to it and how one should be in a local context: "In practical terms, place narratives inform understandings of what kind of place this is, what kind of people are living here and how people like 'us' live" (Junnilainen 2020, p. 46). In line with this, Sofya Aptekar, who has studied how urban dwellers experience and understand their neighbourhoods, argues that "collective representations of place" can shape the ways people imagine the present situation and the future development of that place, turning placed-based narratives and collective imaginings of place into relevant objects of research for studying societal change (Aptekar 2017).

As mayors are so closely related to place-based politics, as well as being the official representative of the place-based official entity, we expect them to be both important (re)producers of place-making narratives and highly relevant sources for identifying these narratives.

In the following, we will present three collective narratives about immigration and immigrant integration that we identified in the interviews with the 13 mayors. Intrigued by these narratives, we conclude by discussing the ways in which the mayors engage in wider debates about rural authenticity.

"It is good for us": a narrative of positive immigration and the resourceful immigrants

When analysing the interviews, an overarching finding is that the mayors' reflections about immigration and integration are clearly contextual. The mayors focus on the specific municipality's situation at a specific time and during its relatively recent history, thus making them varied and diverse. Still, the interviews also showed several similarities. One striking similarity was the collective narrative that we have labelled "It is good for us". This narrative was communicated both in an implicit manner, such as through an obvious sense of satisfaction permeating the interview, as well as in an explicit and condensed manner, via sentences such as "immigration has been good for us" and "immigration has been very positive for this municipality".

When examining this collective narrative about good experiences more closely, we found that it typically revolved around three aspects: demography, competence and cultural diversity. Whereas some of the mayors referred to all three more or less equally, others had a stronger emphasis on one or two aspects. In most interviews, however, all were mentioned several times.

For the aspect of demography, we refer to the notion that immigration is contributing positively to local population growth. While all mayors presented this as a positive effect, the mayors of the municipalities with the smallest populations framed this as particularly important. This was also the case for municipalities where some districts had been suffering from depopulation. In such places, the immigrants had countered this tendency toward depopulation and lit up windows that had been dark for years, refurbished dilapidated houses, rescued the local shops and performed other restorative acts of great importance to the local communities.

The second aspect, competence, is found in how the mayors talk about the resources immigrants bring with them to the community. This includes educational background and professional competence, as well as less formal skills, such as athletic talents and interesting hobbies that have a positive effect on everyday life in the local community. For example, one mayor said that the average level of education in her municipality had increased significantly due to immigration, and s/he was proud to tell us that the municipality at the time of the interview was home to "three professors, medical doctors, butlers, accountants, car mechanics, as well as jewellers" (Mayor 3). Other mayors shared stories about immigrants who had made a huge effort in volunteer organisations or sports clubs.

The third aspect we identified is what we have labelled diversity, which refers to how getting to know people from other parts of the world with different cultures is presented as enriching for the local community. This interpretation of diversity refers to ways of being (since immigrants were often described as more social and outgoing), as well as ways of doing (their tastes in clothing, food, music, etc.). In some interviews, diversity seemed to be viewed as a refreshing or exciting contrast to the homogeneity of "authentic rural life" (Haugen and Villa 2016). This is exemplified by a quotation from the interview with the mayor of one of the biggest and most densely populated municipalities in our study:

> We have always settled more refugees than we have been asked to. Because we see it as very valuable, both for the Norwegian society and for us as a municipality. [...] Immigration makes the inhabitants more diverse. And that is a value in itself. That we don't get this "mono-culture". Diversity in itself is a value. It also makes us more enlightened. We actually learn more about other people and other cultures. I think this is the most important value.
>
> (Mayor 4)

This quote points to the discourse of diversity that has also become the dominant rhetoric in rural areas (Gullikstad and Kristensen 2021). In this context, diversity is presented as a needed and valued learning process for the local inhabitants, who thus come across as somewhat homogenous, but also as open to what can be labelled rural cosmopolitanism (Woods 2018; Villa 2019).

Another example of how this collective narrative of positive immigration and resourceful immigrants was expressed, which touches upon both competence and diversity, can be found in the interview with the mayor of one of the smaller municipalities included in the study. In this municipality there had been labour migrants for many years, and since 2000 quite a few refugees had settled there. Positive experiences were mentioned by the mayor several times during the interview, and it was made explicit when the mayor was asked about the effect immigration has had on the municipality:

> I definitely think of the immigrants as a positive contribution to this municipality, because we need people to work in the local industry. In addition, I think it is positive that they are bringing with them some new cultures. They give us inspiration, both when it comes to food and when it comes to ways of being. Some of them, such as the Eritreans, are very social. And, as I see it, this is very good for this municipality. [...] So, I definitely see the immigrants as a resource to us.
>
> (Mayor 2)

In this excerpt, which undoubtedly can be described as positive, the mayor is quite specific about the advantages s/he thinks come with immigration to this municipality, namely workforce expansion and cultural diversity. S/he is also conveying an understanding of the immigrants as resourceful and as carriers of differences that influence positive changes in the municipality.

"It has worked out very well": a narrative of successful integration

The second collective narrative we have identified is that of successful integration. Whereas the narrative about positive immigration and resourceful immigrants seem to relate to all kinds of migrants, the narrative about integration is focused mostly on refugee settlement, which is not surprising since this is what the Norwegian integration policies are directed towards (Djuve and Kavli 2019). What is somewhat surprising, though, is the positivity the mayors display when talking about integration, since a rather prominent discourse on the national level is that integrating (non-Western) refugees has proven difficult, and that some of the integration policies and measures have failed (Djuve and Kavli 2019). This does not mean, however, that concerns were not raised by the mayors. What it does mean is that, when it comes to the issue of integration, the main message the mayors conveyed is that of a successful municipality.

A telling example of how this narrative was captured is to be found in the interview with the mayor of one of the smaller municipalities included in this study. This municipality has a long history with labour migrants, and ten years ago they also started receiving refugees. The decision to settle refugees was a response to a call from the national immigration directorate of integration and diversity, which according to the mayor was highly contentious: "The idea of settling refugees was very strange and unthinkable for this municipality. And there were also many local politicians who protested" (Mayor 1). During these ten years, the municipality has received a substantial number of refugees, which has had positive repercussions for the local economy and the labour market. At the time of the interview, the mayor informed us that due to the decreasing number of refugees coming to Norway, the plan was to downsize the local integration office and the adult education establishment. When asked about the municipality's experiences with immigration, the mayor gave this answer:

> This municipality's experience with integration is that it is possible to succeed if you invest enough resources, enough energy and get very clever people who are dedicated to the task and really go all in. The people being employed in the integration office need to be of unimpeachable integrity and very competent. In addition, we have an advantage as the local industry is in constant need of labour. This makes it more attractive [both to come and to settle], and it is possible to make special arrangements for training and qualification. And, in this area we have succeeded. One of the challenges we see, however, is that the industry is located on the outskirts of the municipality, and the distances can be rather long, and then it can be difficult for someone to actually get there.
>
> (Mayor 1)

In this excerpt, which clearly revolves around successful integration, the mayor highlights the benefits of this particular municipality, whilst also admitting not everything is ideal.

In addition to organisational and material specificities, the narrative of success also includes stories about the open-mindedness and hospitality of the municipalities' locals, which does not apply for everyone, but for enough to make integration possible. In other interviews, the local advantages are related to the geographical localisation of the municipality or its size. For example, the mayor of the smallest municipality included in this research said that the local politicians were disappointed when they received the news that the municipality would not receive any more refugees due to national regulations, adding:

> I think that integration is easier in small municipalities than in big. I have heard this from the refugees themselves, too. [It is] because they have friends living in other municipalities and they see that living in a bigger city is very different from living here.
>
> (Mayor 3)

Other mayors pointed to historical facts and past experiences when explaining why integration has been so successful in their particular municipality. The mayor of one of the coastal municipalities remarked:

> Our experience is that it is easier for us [as opposed to municipalities in the inland], because we are more used to being in contact with strangers. It has been labelled as part of our culture, some kind of coastal hospitality.
>
> (Mayor 13)

Others referred to previous experiences with diversity, which were presented as a local competence that made this particular municipality especially suited for successful immigrant integration. This was the case with the mayor of an inland municipality, who also told us that most refugees being settled in the municipality would move as soon as the system would let them, as they could not get a job there:

> Previously, there was a big institution for mentally disabled people in this municipality. They came from all over Norway and lived in this local community. So, in my opinion, we have a history where we have been good at integration. Of all kinds of people. Of course, there are also some aspects related to this institution that we are not so proud of, but still… At least we are used to diversity even though we have not had many immigrants living here. […] I have chosen to think that we have been very good at seeing different kinds of people. No matter where you come from, if you have a job or not. Everyone is important, no matter who you are.
>
> (Mayor 7)

By drawing on these previous experiences with diversity, this mayor makes the same point as the other mayors – namely, that the municipality's experiences with immigrant integration can be described as a success. The underlying message conveyed in all these stories, moreover, is that integration is difficult and that to succeed you need to be good and do the right things.

"There will always be someone". A narrative of handling normalised incidents of xenophobia and opposition to immigration

In parallel to the many nice and proud stories of resourceful immigrants and well-functioning municipalities with openminded and tolerant locals, the mayors spoke of the other side of the coin. They told us about various experiences with xenophobia and hostility towards immigrants, as well as opposition to immigration more generally or immigration to this municipality specifically. Even though these experiences were presented as negative in the sense that they represented attitudes and values that were seen as unacceptable, the main message presented in these stories is that these attitudes were not representative of this municipality. They were instead a marginal phenomenon that was followed up on when needed. In other words, the collective narrative in these stories was not of failure and conflict, but rather yet another story of local communities and politicians that are up to date and under control.

One very explicit way of talking about such incidents of xenophobia which clearly demonstrates both normalisation and particularisation, as well as the political willingness and ability to act upon this, is Mayor 2's story about an open meeting s/he had recently attended on the very outskirts of the small municipality. During the meeting, one of the locals stood up and complained that he felt he was "in the centre of Mogadishu" when visiting a shop close to his home a few days earlier. Upon hearing this, the mayor phoned someone to follow up, only to hear that the adult education establishment had visited the local shop that particular day on their way to one of the local tourist attractions. As the mayor recounted, "They came in a bus, probably 25 people, and I guess that felt somewhat overwhelming in this local shop. But no, I don't feel there has been any xenophobia" (Mayor 2).

By speaking about this somewhat funny incident on the periphery of this small municipality, the mayor is both situating xenophobic attitudes away from the dominant place-based culture and thereby making them marginal and harmless, and positioning him/herself as someone who pays close attention to this kind of unwanted attitude or statement, hence preventing it from getting out of control.

A similar story is found in the interview with the mayor of another small municipality, one which has a long history of labour migration but a shorter history of settling refugees. This mayor explained that while most long-term residents do not have any objections toward the immigrants living there, there were "four or five people" who "occasionally" write negative remarks on social media. However, according to the mayor, these people "live in the same place and we all know who they are" (Mayor 8). By stating his personal knowledge of these people and the attitudes they are promoting, the mayor is not only marginalising their attitudes, but also implying that they are harmless. Just as did Mayor 2, moreover, s/he demonstrates that the situation has been followed up on and as such is under control.

Concluding reflections

The collective narratives we have outlined in this chapter describe the ways in which small municipalities in a rural region of Norway have succeeded in the challenging task of hosting various kinds of immigrants, including them in the local community and avoiding the kind of racialised conflicts that have evolved in many other places both in Norway and elsewhere.

As politicians, mayors are engaging in impression management in presenting their municipality in an advantageous light. This means that their stories can be seen as positioned voices in a political terrain saturated with discussions about what kind of communities are most attractive, which communities should be awarded public support and which communities will continue to thrive in the future. By applying a narrative perspective, we have approached these stories as sense-making activities that provide information about both individual meaning-making and the wider social context in which this meaning-making takes place. Considering narratives as historically informed, collective processes of place-making that affect what defines the community and what does not, as well as local self-understandings and identities, we argue that setting these narratives in dialogue with the research we present in the chapter's introduction offers new insights about the position of the rural, rural authenticity and rural identity in both a national and an international context.

The first discussion we will engage in is that of rural societies' motivation for welcoming new immigrants. First of all, the narratives we have identified counter the notion that rural areas and small municipalities welcome new inhabitants because they contribute to the maintenance or even regeneration of the communities. Even though the mayors' stories show that immigrants do have a positive effect on both local demography and the local industry, they also convincingly demonstrate that the positive contribution is not only about demographic and economic growth, but also about becoming more culturally diverse, getting to know new people and cultures, and being open to change.

By highlighting this wish for more diversity, on the one hand the narratives are reproducing the well-known notion of the rural as typically homogeneous, "native" and "rooted". On the other hand, though, as these characteristics are presented as not ideal, the narratives are also giving new meanings to the notion of the rural, such as heterogenity and cosmopolitanism. In line with this, the narratives challenge the claim that immigrant integration is particularly difficult in rural societies. By sharing their experiences of xenophobia and opposition towards immigrants, however, the mayors acknowledge that these negative tendencies do exist. At the same time, by demonstrating the ways in which these "unwanted" tendencies are under constant surveillance, which is doable in small communities characterised by transparency, the narratives also paint a picture of a community where everyone, including international immigrants, can feel secure. The open reflections upon the xenophobia and opposition, moreover, point to a community wherein emotional worries about cultural exchange are openly discussed.

To sum up, the overarching message conveyed in these three narratives is an understanding of the municipalities as generous and openminded. This self-understanding is, in part, related to a localised historical authenticity about mobility and openness, and to the skills needed for surviving as small, sparsely populated communities and official entities with some kind of local autonomy in a context where everything small is constantly under pressure of being merged into other, larger entities. The narratives can therefore be described as a kind of "narrative attractiveness" (Lysgård and Cruickshank 2016), which alludes to some kind of competition between local communities and municipalities in terms of being desirable in order to attract future inhabitants and keep those who have already come to stay.

As such, the local, rural authenticity the mayors draw on and (re)produce in our interviews was not about defending traditions and fighting change, but rather about being able to adapt to new situations and having the courage to grab the possibilities that come along even though the outcome is uncertain – in other words, to take an active part in the new, globalised, modern world.

Notes

1 *Living Integration: At the Crossroads between Official Policies, Public Discourses, and Everyday Practices*, funded by the Norwegian Research Council (261982).
2 The interviews were conducted in Norwegian.

References

Aasbrenn, K. and Sørlie, K. (2016). "Uttynningssamfunnet – 25 år etter". In Villa, M. and Haugen, M. S. (eds.), *Lokalsamfunn*. Oslo: Cappelen Damm Akademisk, pp. 152–176.

Åberg, M. and Högman, A. K. (2015). "Histories meet histories: A pilot study of migration and civil society in Swedish medium-sized cities, small towns and villages". *Journal of Civil Society*, 11(2), pp. 187–203.

Almås, R., Haugen, M. S., Rye, J. F. and Villa, M. (2008). "Omstridde bygder". In Almås, R., Haugen, M. S., Rye, J. F. and Villa, M. (eds.), *Den nye bygda*. Trondheim: Tapir akademisk forlag, pp. 9–28.

Amundsen, B. (2017). " Innvandrere skaper nytt liv i bygde-Norge". *Forskning.no*, 3 July. Available at: https://forskning.no/arbeid-innvandring-norges-forskningsrad/innvandrere-skaper-nytt-liv-i-bygde-norge/336411 (Accessed 30 September 2020).

Aptekar, S. (2017). "Looking forward, looking back: Memory and neigborhood identity in two urban parks". *Symbolic Interaction*, 40 (1), pp. 101–121.

Arora-Jonsson, S. (2017). "Development and integration at a crossroad: Culture, race and ethnicity in rural Sweden". *Environment and Planning*, 49 (7), pp. 1594–1612.

Aure, M., Førde, A. and Magnussen, T. (2018). "Will migrant workers rescue rural regions? Challenges of creating stability through mobility". *Journal of Rural Studies*, 60, pp. 52–59.

Befring, Å. M. (2017). "Frykter for at de ikke får flyktninger". *NRK.no*, 28 October. Available at: www.nrk.no/norge/xl/na-er-det-en-kamp-om-a-fa-flyktninger-1.13753780 (Accessed 30 September 2020).

Berg, N. G. and Lysgård, H. K. (2004). "Ruralitet og urbanitet – byd og by". In Berg, N. G., Dahle, B., Lysgård, H. K. and Løfgren, A. (eds.), *Mennesker, steder og regionale endringer*. Trondheim: Tapir akademisk forlag, pp. 61–76.

Berg-Nordlie, M. (2018). "New in town: Small-town media discourses on immigrants and immigration". *Journal of Rural Studies*, 64, pp. 210–219.

Bo, I. G., Christensen, A. D. and Thomsen, T. L. (eds.) (2016). "Narrativ forskning: Tilgange og metoder". In *Narrativ forskning. Tilgange og metoder*. København: Hans Reitzels forlag, pp. 13–34.

Bygnes, S. (2017). "Welcome to Norway! Da 'flyktningkrisa' kom til Norge". *Tidsskrift for velferdsforskning*, 20 (4), pp. 286–301.

Djuve, A. B. and Kavli, H. C. (2019). "Refugee integration policy the Norwegian way: Why good ideas fail and bad ideas prevail". *Transfer*, 25 (1), pp. 25–42.

Finney, N. and Robinson, V. (2009). "Local press, dispersal and community in construction of asylum debates". *Social & Cultural Geography*, 9 (4), pp. 397–413.

Gullikstad, B. and Kristensen, G. K. (2021). "Vi er et mangfoldssamfunn'. Mangfold som offentlig integreringsfortelling: (Re)produksjon av likhet og ulikhet?" In Gullikstad, B., Kristensen, G. K. and Sætermo, T. (eds.), *Fortellinger om integrering i norske lokalsamfunn*. Oslo: Universitetsforlaget, pp. 40–63.

Hadj Aboud, L. (2019). "Immigrant integration: The governance of ethno-cultural differences". *Comparative Migration Studies*, 7 (1), pp. 1–8.

Haugen, M. S. and Lysgård, H. K. (2006). "Discourses of rurality in a Norwegian context". *Norwegian Journal of Geography*, 60 (3), pp. 174–178.

Haugen, M. S. and Villa, M. (2006). "Big brother in rural societies: Youths' discourses on gossip". *Norwegian Journal of Geography*, 60 (3), pp. 209–216.

Haugen, M. S. and Villa, M. (2016). "Lokalsamfunn i perspektiv". In Villa, M. and. Haugen, M. S. (eds.), *Lokalsamfunn*. Oslo: Cappelen Damm Akademisk, pp. 17–33.

Hedberg, C. and Haandrikman, K. (2014). "Repopulation of the Swedish countryside: Globalisation by international migration". *Journal of Rural Studies*, 34, pp. 128–138.

Hubbard, P. (2005). "Inappropriate and incongruous": Opposition to asylum centres in the English countryside. *Journal of Rural Studies*, 21(1), pp. 3–17.

IMDi. (2020). *Tall og statistikk*. Available at www.imdi.no/om-integrering-i-norge/ (Accessed 30 September 2020).

Jentsch, B. (2007). "Migrant integration in rural and urban areas of new settlement countries: Thematic introduction". *International Journal on Multicultural Societies*, 9 (1), pp. 1–12.

Junnilainen, L. (2020). "Place narratives and the experience of class: Comparing collective destigmatization strategies in two social housing neighborhoods". *Social Inclusion*, 8 (1), pp. 44–54.

Karlsdóttir, A., Sigurjónsdóttir, H. J., Heleniak, T. and Cuadrado, A. (2020). *Learning to live in a new country – everyday social integration. Civil society and integration – Nordic rural perspectives*. Nordic Council of Ministers.

Kasimis, C., Papadopoulos, A. G. and Pappas, C. (2010). "Gaining from rural migrants: Migrant employment strategies and socioeconomic implications for rural labour markets". *Sociologia Ruralis*, 50 (3), pp. 258–276.

Kristensen, G. K. and Sætermo, T. (2021). "Hvordan lykkes med integrering? Stedsproduserende fortellinger om integreringsarbeid i to rurale kommuner". In Gullikstad, B., Kristensen, G. K. and Sætermo, T. (eds.) *Fortellinger om integrering i norske lokalsamfunn*. Oslo: Universitetsforlaget, pp. 109–133.

Lysgård, H. K. and Cruickshank, J. (2016). "Attraktive lokalsamfunn – hva er det for hvem?" In Villa, M. and Haugen, M. S. (eds.), *Lokalsamfunn*. Oslo: Cappelen Damm Akademisk, pp. 95–114.

McAreavey, R. and Argent, N. (2018). "New Immigration Destinations (NID): Unravelling the challenges and opportunities for migrants and for host communities". *Journal of Rural Studies*, 64(1), pp. 148–152.

Meissner, F. and Heil, T. (2020). "Deromanticising integration: On the importance of convivial disintegration". *Migration Studies*, mnz056, 13 February, pp. 1–19.

Milbourne, P. and Kitchen, L. (2014). "Rural mobilities: Connecting movement and fixity in rural places". *Journal of Rural Studies*, 34, pp. 326–336.

Neal, S. (2002). "Rural landscapes, representations and racism: Examining multicultural citizenship and policy-making in the English countryside". *Ethnic and Racial Studies*, 25, pp. 442–461.

Nicolajsen, S. (2019). "Pause ikke nok". *Klassekampen*, 20 February, https://klassekampen.no/utgave/2019-02-20/pause-ikke-nok (Accessed 30 September 2020).

Oliva, J. (2010). "Rural melting-pots, mobilities and fragilities: Reflections on the Spanish case". *Sociologica Ruralis*, 50 (3), pp. 277–295.

Ray, L. and Reed, J. K. (2005). "Community, mobility and racism in a semi-rural area – comparing minority experience in east Kent". *Ethnic and Racial Studies*, 28 (2), pp. 212–234.

Riessman, C. K. (2008). *Narrative methods for the human sciences*. Los Angeles: Sage Publications.

Rye, J. F. (2014). "The Western European countryside from an Eastern European perspective: Case of migrant workers in Norwegian agriculture". *European Countryside*, 6 (4), pp. 327–346.

Rye, J. F. (2018). "Labour migrants and rural change: The 'mobility transformation' of Hitra/Frøya, Norway, 2005–2015". *Journal of Rural Studies*, 64, pp. 189–199.

Rye, J. F. and Scott, S. (2018). "International labour migration and food production in rural Europe: A review of the evidence". *Sociologica Ruralis*, 58(4), pp. 928–952.

Søholt, S. and Aasland, A. (2019). "Enhanced local-level willingness and ability to settle refugees: Decentralization and local responses to the refugee crisis in Norway". *Journal of Urban Affairs*, 11 February. https://doi.org/10.1080/07352166.2019.1569465.

Søholt, S., Stenbacka, S., and Nørgaard, H. (2018). "Conditioned receptiveness: Nordic rural elite perceptions of immigrant contributions to local resilience". *Journal of Rural Studies*, 64, pp. 220–229.

Søholt, S., Tronstad, K. R., and Bjørnsen, H. M. (2014). *Innvandrere og sysselsetting i et regionalt perspektiv. En kunnskapsoppsummering* (vol. 25), Oslo: NIBR, HiOA.

Steen, A. (2009). "Hvorfor tar kommunene imot 'de fremmede'? Eliter og lokal skepsis". In Saglie, J. (ed.), *Det Nære Demokratiet*. Oslo: Abstrakt Forlag, pp. 323–350.

Steen, A. and Røed, M. (2018). "State governance or local agency? Determining refugee settlement in Norwegian municipalities". *Scandinavian Journal of Public Administration*, 22(1), pp. 27–52.

Svensson, T. (2019). "Landets beste på integrering?" *Agenda magasin*, 12 July https://agendamagasin.no/artikler/landets-beste-pa-integrering/ (Accessed 30 September 2020).

Tønnesen, M. and Andersen, S. N. (2019). *Bosettingskommune og integrering blant voksne flyktninger. Hvem bosettes hvor, og hva er sammenhengen mellom bosettingskommunens egenskaper og videre integreringsutfall?* Oslo: SSB 2019/13.

Valenta, M. (2007). "Daily life and social integration of immigrants in city and small town: Evidence from Norway". *International Journal on Multicultural Societies (IJMS)*, 9 (2), pp. 284–306.

Villa, M. (2019). "Local ambivalence to diverse mobilities: The case of a Norwegian rural village". *Sociologia Ruralis*, 59 (4), pp. 701–717.

Woods, M. (2011). *Rural*. Oxon: Routledge.

Woods, M. (2012). "New directions in rural studies?" *Journal of Rural Studies*, 28 (1), pp. 1–4.

Woods, M. (2018). "Precarious rural cosmopolitanism: Negotiating globalization, migration and diversity in Irish small towns". *Journal of Rural Studies*, 64, pp. 164–176.

12 Dynamics of changes in the farmers' contestation in Poland in 1989–2018

On the way to rationality and an institutionalised model of collaboration

Grzegorz Foryś

Introduction

Contestation activities of peasants and farmers have a long tradition in Europe generally, and Poland specifically, going back to the Middle Ages. Before World War II, peasants represented the largest component of the social structure in many European countries. In Poland, this was the case until the 1950s. In the past, Polish society was considered a "peasant society" (Wasilewski, 1986), and that was not unjustified. Today, this is clearly not about the numerical dominance of the rural population in the social structure, and also not about the domination of farmers in professional categories. Nowadays, references to "peasant society" point to the peasant roots of many individuals and groups constituting Polish society and how these roots shape its image.

At the beginning of the 1990s, radical political, economic and social changes began to occur in Central and Eastern Europe. This trend was strong in Poland, affecting farmers and rural areas right from the start. Throughout the first decade of the transformation, the incomes of farm owners were reduced to the greatest degree in comparison with the incomes of other professional categories. The intensity of the restructuring processes, as well as concurrent modernisation and the introduction of a market-oriented economy, led to a decline in the number of agricultural farms in Poland.[1] This trend mostly brought about the liquidation of the smallest farms, which were only loosely connected with the market and not able to keep up with the free market competition. The owners of the largest farms had to face the following challenges: changeability of prices of agricultural products, lack of profitability of production, expensive loans, lack of funds for modernisation and an influx of agricultural products from Western countries. All of the above factors contributed to preparing the ground for farmers' contesting activities.

The main goal of this chapter is to examine the degree to which farmers' protests in Poland over the last 30 years could be seen as inscribed in the model of interest representation and perceived as organised cooperation, and the extent to which this type of relationship between farmers and the state is reflected in the awareness of protesting farmers, In my view, the main tendencies of farmers' protests in Poland include professionalisation, routinisation

DOI: 10.4324/9781003091714-14

and institutionalisation. Consequently, many aspects of farmers' contestation activities in Poland have been quite similar to those conducted by farmers in Western Europe. In farmers' relations with the state, the role of protests is declining while the rising importance of a permanent, institutionalised cooperation between the state and farmers can be observed, and this is reflected in binding legislative solutions. Protest becomes a type of action that is somewhat auxiliary to activities conducted in the institutional sphere. This is reflected in farmers' consciousness and collective imagination influencing their perception of the various methods and solutions that protect their interests. In their opinion, the most effective kinds of contesting are activities based on permanent organisational forms (producer groups, trade unions and interest groups) and not direct protest actions. In other words, they diagnose their interests, as well as the ways in which they are implemented, in an authentic and pragmatic way. This also means moving away from the traditional peasant symbolism that accompanies protests while preserving elements of populism in the symbolic sphere.

The analysis will cover three main threads. In the first, I point out the main processes which have been the subject of agricultural protests in Poland over the last 30 years. In the next thread, I refer to quantitative data and present farmers' opinions on the forms of defending their own interests. In the last, I present comments on the model of organised cooperation in Polish agriculture in the context of its relations with the populist symbolism of protests.

The analyses presented here will refer to data compiled from 1989 to 1993 and published by Grzegorz Ekiert and Jan Kubik, who collected them within the framework of an international project describing protests in Poland, Slovakia, Czech Republic, Hungary and East Germany, as well data from 1997–2001 pertaining to on my own research on peasants protests that I (Foryś, 2008) conducted with the use of protest events analysis as a research method (Tilly, 1995). I have collected more recent data in a less systematic manner through analysing the selected protest events of the greatest footprint and range that were reported by the press, online news media and the largest Internet portals. In addition, I will use data from a nationwide quantitative survey conducted on a sample of 3,551 farm owners.[2]

Evolution of farmers' protests in Poland from 1989 to 2018

In more than a decade of transformation of the Polish political, economic and social system, in the years 1989–2001 two waves of farmers' protests were noted. The first wave encompassed the timeframe from 1989 to 1993, with 112 protest events taking place. The second wave lasted from 1997 to 2001, with 58 protest events documented. The "term protest" event refers to the tradition of the analysis of protest events method which, despite criticism, provides an opportunity to record qualitative and quantitative aspects of protests (Koopmans & Rucht, 1999). As stated by Grzegorz Ekiert and Jan Kubik: "analysis of events is a particularly useful tool applied to construct the most voluminous and systematic data sets related to protest actions and behaviors, as well as their various components and dimensions" (Ekiert & Kubik, 1997, p. 26).

To gain a better understanding of the characteristics of protests, it is worth exploring what a "protest event" really means. A protest event can be understood in several different ways: it may be defined as an individual action of a protest character, as well as a more complex type of action, such as a series of protests (with protests occurring in many places at the same time sharing identical or similar demands) and protest campaigns (sequence of protests, which are planned and organised by one coordination centre). This also means that terms such as "series of protests" or "protest campaign" might include tens or even hundreds of protest actions conducted at the same time in various places. This was the case with the two analysed waves of farmers' protests in Poland, as illustrated by the fact that 53 protest events were conducted at the end of 1990, and the beginning of 1991 saw more than 500 connected protest actions. To compare, in the final months of 1998 and the beginning of 1999 there were 230 such events. More crucial was participation in specific types of protest actions. While in the entire analysed time frame the prevalence of individual protests was noticeable, the share of protests involving identifiable protest series and campaigns increased from the first to the second protest waves, from 24.2% to 40.3%. To some extent, this was indicative of the cooperation development between agricultural organisations that, at the time, dominated as protest organisers, namely NSZZ RI Solidarność, ZZ Samoobrona and KZRKiOR.[3] This was confirmed by declining number of spontaneous actions, to the benefit of actions organised by the unions mentioned above and their centres of coordination. The average shares of spontaneous action in each wave of farmers' protests were 31.2% and 11.1%, respectively.

Despite this trend, which indicated the increasing organisational potential of the protesters, other indicators also deserve some attention as shortcomings of farmers' mobilisation. First, protests engaging fewer than 1,000 people were predominant and, on average, during the two waves of protests they comprised 32.9% and 70.5%, respectively, of all protest activities. Actions with 10,000 or more participants were noticeably less frequent, on average constituting 9.2% of the first wave of protests and 19.0% of the second wave. A similar tendency can be seen in the data on the ranges of protests. During both waves of farmers' protests, those of a local character dominated, with a share of 59.1% for the first and 73% for the second wave of protests, while the percentage of national protest reached 25% and 19.3%, respectively. What most distinguished farmers' protests in Poland was their economic aspect. In both the first and second wave of their contesting actions, farmers mostly formulated material demands (such as credits and loans subsidies, agricultural fuels or prices of agricultural products) and demands for economic changes (for example, establishing agricultural agencies, protection of agricultural markets, suitable conditions for stabilisation of production and better profitability over a long period of time). The joint share of both categories of economic demands remained at around the same level during both waves, with 82.5% and 79.5% respectively.

Closer analysis of economic demands, formulated over a decade of protests in the new free market and democratic reality, revealed an interesting tendency for both subcategories of these type of demands. As noted, they had to do with either economic changes or material compensation. This meant a decreasing share of the former and a growing percentage of the latter. During the first wave of protests, demands for economic changes were 64.3%, and at the end of the second wave they went up to 80%. Conversely, demands for material compensations were 32.1% and 13.3%, respectively. This might serve as evidence of the maturing consciousness of farmers and their willingness to undergo deep changes of a complex, sometimes even an institutionalised, character, and consent to adjust to the challenges of the free market. This could be explained by the fact that the protest participants were owners of large, already modernised farms or of farms undergoing modernisation. In other words, farmers understood that the compensatory actions are insufficient and had short-term effects. They should be replaced by comprehensive reforming actions.

From the point of view of interest representation, the general tendencies observed in the following years appear to be crucial as they acquired a more permanent character and also contributed to the development of the organised collaboration system. First, farmers' protests went through the process of professionalisation, which was reflected in an increased percentage of campaigns and protest series. This was closely connected with more intensive cooperation between farmers' trade unions. Second, throughout both waves of protests, the institutionalisation of protest actions could be seen in the establishment of permanent cooperation between the centres of farmers' unions. This was particularly the case during the second wave of protests, and was in clear contrast with the competition between them at the beginning of the 1990s. Third, gradual routinisation of farmers' protests was one of the signs of protests' institutionalisation. This meant applying certain patterns of organisational activities, and especially the protest methods, which evolved with time and acquired a specific repertoire (road blockages, building occupation, destruction of crops). Fourth, the first wave of protests attracted various categories of farmers, especially in its early phase when initial reforms were being implemented. In the second wave, only owners of large, specialised agricultural farms participated in the protests. This revealed the differences in interests among farmers: not just the internal division of this category into owners of small and large farms, but also the differentiation of interests between various groups of agricultural producers. Although all categories of farmers had reasons for protesting, not all of them had sufficient mobilisation potential or organisational background. Owners of small farms, due to weak links with the market, lack of participation in producer associations and territorial dispersion, lacked such capacity. Moreover, each group of agricultural producers only cared about the interest of its own environment, and did not engage in the protest activities of other categories of producers, which, to some extent, suggested a lack of solidarity.

These tendencies developed over the first 15 years of transformation and strengthened in the years that followed. For many reasons, the most crucial, breakthrough moment came with Poland's accession to the European Union (EU). Initially, prospective EU membership provoked intensive resistance by farmers as they were fearful of competition from Western European farmers. Over time, it became clear that EU membership brought farmers significant advantages, making them the main beneficiaries of EU membership and resulting in accelerated modernisation of agriculture in Poland. This was mainly due to the fact that Poland's membership in the EU took place after changes in the structure of Polish agriculture and the disappearance of about 30% of farms, which did not stop the market competition.

It should also be noted that farmers' protests were going through a process that could be described as "Europeanisation". This new form of protest, which Sidney Tarrow termed "Euro protest" (Imig & Tarrow, 2001, p. 32), consisted in the participation of representatives of at least one EU member state in a protest in which demands were made against European institutions. Such protests among Polish farmers were quite frequent. Poles were among the dairy farmers and wheat producers who protested in Brussels in 2008. In 2012, dairy farmers protested against the liquidation of milk quotas. Two years later, wheat producers from Poland protested against low compensation for losses caused by the embargo on Russia. In 2015, Polish dairy farmers also participated in farmers' protests in Brussels. Farmers engaged in the production of beef, pork, fruit and vegetables also protested. A year later, the protests of wheat producers were again apparent.

Protests at the national level took place concurrently with those in Brussels and also at various times before Poland's EU accession. One of the most notable protests took place in 2003, when farmers, prompted by declining incomes, clashed with police. Dramatic events occurred mostly in Greater Poland (Wielkopolska), known for its large, modern farms, and resulted in several arrests and farmers being injured due to use of rubber bullets by police. Pork producers protested in 2007, followed by wheat producers in 2008 and 2009. In 2013, farmers demonstrated against the sale of agricultural land to farmers from Western Europe and the generally unsatisfying agricultural policy. The continuation of protests against these sales of farmland could be seen in 2015, and new reasons for protests also emerged, including calls for increased profitability of milk and pork production, compensation for agricultural losses caused by boars and bans on the ritual slaughtering of animals. The year 2017 was marked by protests by poultry and fur animal farmers, while in 2018 protests by potato (and other vegetable) farmers were notable.

It is worth adding here that a perpetual element of the agricultural protests was the symbolism used by the protesters. Its status remained unchanged in the following decades, but farmers' attitude to it evolved. The symbolism to which they referred, as well as the values behind it, was the basis for mobilising their actions and for building their common identity. The collection of symbols used by protesting farmers referenced three areas. The first was traditional cultural symbols of Polish society (patriotic songs, national flags, religious symbols); the second was values related to the peasant ethos (references to land and farm work; scythes, sickles and claws as traditional tools of peasant work; slogans referring to the misery of the peasant in Polish history); and the third area was related to the experiences of transformation and was formed under the influence of direct events in which the farmers were involved (burning effigies of politicians, destruction of agricultural crops). At this point we are most interested in the second area referring directly to the peasant ethos. Its changing presence in the protests over three decades is indicative of changes in farmers' awareness. While this kind of symbolism dominated in the last decade of the 20th century, in the 21st century, especially after Poland's accession to the EU, it played a marginal role, before disappearing completely in the last decade. The protesters eventually resigned from peasant self-identification, which they began to consider as offensive to them as agricultural entrepreneurs. A key role was played by a meaning framework referring to national symbols, emphasising the equal status of farmers with other interest groups and those resulting from direct experience of fighting for one's own interests, mainly highlighting the current disadvantages of farmers. This trend shows one of the dimensions of rationalisation of the activities of this social category and the process of de-ideologising their interests, which fits into the model of institutionalised cooperation. It should be added, however, that although the real problems of agricultural producers were accompanied by rational demands on their part, there were elements of populism in the symbolic sphere. These were expressed in the main frame of meaning used by the protesters, in which they referred to national symbolism and presented themselves as defenders of Polish land, Polishness and the Polish nation.

In sum, the dynamics of farmers' protests in the last 30 years mostly depended on economic factors. In the first decade of transformation, these included a sudden drop in farm owners' incomes (such that farm owners in 1990–1993 made only 66% of their 1989 incomes), difficulty in getting farm loans and credits, low prices of agricultural products and a lack of customs regulations for imported agricultural products. It can be said that the 1990s were exceptional in terms of economic demands, which were a reaction to the intensification of negative phenomena connected with the economic transformation: the institutional reconstruction of Polish agriculture, the processes of setting modernisation in motion, as well as the emergence of the free market and its further development. Therefore, the extent and intensity of economic influences were greater at that time than in the later years. Obviously, this did not mean that there was a lack of other reasons or causes for protests, such as political issues (e.g., dismissal of a minister or an entire government), changes in agricultural policy or the establishment of various agricultural agencies. However, it should be stressed that political causes were not dominant at any particular point. At best, they accompanied the economic causes.

In the first two decades of the 21st century, the causes of the protests were quite similar. It could be said that a rather stable set of such causes and reasons was established. They usually appeared as a result of the increasing costs of agricultural production, unsatisfactory prices of agricultural products, problems with their wholesale purchases (especially cereals, hogs and milk) and inadequate compensation for losses connected with the Russian embargo or negative consequences of unfavourable weather conditions, such as floods and droughts. Such a state of affairs appeared to be natural and indicative of a certain normalisation in the free market context, yet at the same time ran contrary to the situation of the real socialism era, wherein farmers' protests were mostly determined politically and involved efforts to maintain private ownership of land, resistance against compulsory supplies of agricultural products to the state and defence of farmers' rights and freedoms.

Four models of interest representation

Numerous academic disciplines, especially political science, law and sociology, give attention to the systems of interest representation. The variety of definitions and meanings connected with the phenomenon of interest representation indicates that, for the purposes of this chapter, its understanding should be narrowed. What can be observed here is the set of links between interests and state decision structures, which indicates a political aspect of such representation. Placing emphasis on the political component is characteristic of the mainstream approach to interest representation. The roots of such thinking are described in the classic work of Hanna Pitkin (1967, p. 3) and in the reflections of Philippe C. Schmitter (1979). These authors present an approach to interest representation which is dominant today, but still debated, referring to both civil society and the authority of the state. In this chapter, I adopt an understanding of the interest representation system following Schmitter, who states that:

> A system of interest and/or attitude representation is an institutional solution that combines interests occurring in civil society, a particular modal or ideal-typical institutional agreement for linking the associationally organized interests of civil society with the decisional structures of the state.
>
> (Schmitter, 1979, p. 9)

In Poland it was Jerzy Hausner who proposed a thorough analysis of possible systems of interest representation (Hausner, 1996). According to him, the course of events considering the variety of coexisting factors could, in the first decade of transformation and subsequent years, result in one of the four models of interest representation in Poland (Hausner, 1996, p. 211). The first model involves authoritarian populism and constitutes the worst possible scenario, as it undermines the principles of parliamentary democracy. In this model, the structures of agency interest are undergoing processes of liquidation, and the articulation of claims is mostly conducted through concessions and control of a mass organisation that can mobilise social masses and block interests that are articulated independently. The second model is that of authoritarian corporatism characterised by combining a statist economy with a monopoly on representation, held by some economic organisations. The probability of this model's implementation in the 1990s in Poland depended to a great extent on the character of the economic reforms. The great strides taken by privatisation, a significant reduction of the state-run economy sectors, as well as a noticeable phase-out of state interventionism in the economy prevented this model from further developing. Observing the current activities of the Polish government, which is aiming for renationalisation of some sizeable enterprises from the industrial and service sectors, as well as the current level of state interventionism in the economy, one cannot rule out that the future might bring development of this type of interest representation system, in which social groups direct their activities towards the state and see state activities as their most effective means of interest protection and the best guarantee of their socio-economic positions. Today, an example of a system of this kind is the Chinese model of capitalism.

The third model is about class confrontation and it assumes that social conflicts are being played out according to the lines of class divisions, mainly between capital and labour. This does not have to result in the negation of democracy, but presents class-based political parties as being able to step out of their formerly accepted frames. Emergence of great social differentiation accompanied by financial crisis might become an obstacle which the state might not be able to overcome or overpower. As a result, numerous social groups are deprived of the possibility to articulate their interests and they might have a sense of exclusion. Finally, there is a fourth model of interest representation, which can be described as a model of institutionalised collaboration, wherein social groups and interest organisations are oriented towards cooperation. They are well aware of their interests and needs. They clearly articulate them and are ready to explore their opportunities. As participants in social life, they shape mutual relations, as well as relations with the state, basing them on trust and cooperation. They perceive this model as more advantageous than fighting for domination.

The models described above, as well as observation of socio-political life in Poland, legitimise the formulation of two findings. The first one, of a more general character, states that, in the last 30 years of transformation (1989–2019), the interest representation model in Poland has evolved from authoritarian corporatism during the first phase of political transformation to the institutionalised collaboration model of today. The second, more detailed consideration refers to farmers' protest activities, admitting that farmers, as a social and professional category, have also followed the above while trying to gain influence on decision-making processes. It should be added that farmers' protest activities created a significant impulse for building the interest representation model on the basis of organised collaboration in a general social sense. This could stem from the specifics of the farming profession in Poland. After the change of political regime in Poland, farmers were the first social category that had to face the demands of the free market. After the demise of the socialist economy of the Communist state they became the first owners of their workplaces due to the fact that 75% of the total area of agricultural land in Poland remained in private hands, even during the Communist rule and despite the pressure of the totalitarian state to collectivise agriculture. Such efforts of the state intensified, especially after 1956.

Model of institutionalised collaboration and farmers' consciousness

This chapter will now consider the empirical data taken from a nationwide quantitative survey conducted on a sample of 3,551 farm owners. The questions included inquiries to prompt an evaluation of farmers' mobilisation potential. At the same time, the answers to these questions could be treated as indicative of change in the consciousness of this socio-professional category. This change can be observed in the emergence of the model of collaboration between farmers and the state. Two variables will be considered: opinions on the existence of organisations defending farmers' interests, and farmers' attitudes towards the methods of defending their interests.

The farmers' opinions presented in Table 12.1 are worthy of a closer look and deserve detailed explanation. The left side of the table presents farmers' perceptions of organisations defending their interests. Here, it is possible to refer to the results of various studies from different periods of time. Using them as a basis, it could be said that farmers currently share a degree of scepticism even about the very existence of such organisations, with only 12.6% of respondents recognising that such organisation existed. This is the lowest in the entire period of all editions of the study, going back to 1994. The highest value was noted in 1999, which was not coincidental: this was the peak of the second wave of protests and, at the time, the main defenders of farmers' interests were the unions (NSZZ RI Solidarność, Samoobrona and KZRKiOR). The number of protest actions and the radical character of these actions, especially those conducted by Samoobrona, left an imprint on the consciousness of farmers, and society as a whole. Additionally, towards the end of the 20th century there were not many farmers' organisations and the few that existed were consolidated. This situation has changed over the last 15 years, as farmers' unions have weakened and the number of producer organisations and industry organisations has increased. This organisational pluralism and differentiation of farmers' interests do not go together with the defence of the latter; in any event, the spectacular forms of defence with numerous protest actions such as those taking place in the 1990s seem to belong to the past. Consequently, farmers have become less inclined to see currently existing farmers' organisations and associations as defenders of their interests. Interestingly, the main variable which differentiated opinion on this matter is farm area/size. In the group of the largest farm owners, only 22% of respondents saw such organisations as their allies. This is no coincidence, as owners of the largest farms are the most likely to be members of producers' associations. Positive attitudes towards farmers' organisations are

Table 12.1 Opinions on the existence of organisations defending farmers' interests and the most effective forms of defence of farmers' interests (in %)

	Are there any organisations defending farmers' interests (all respondents)		Road blockages, demonstrations, occupations	Discussions and pressure on politicians	Organising associations and cooperatives	Does not make sense to do anything	Total
			The most effective forms of fighting for farmers interests (all respondents)				
	Yes	No					
2018	12.6	87.4	10.4	16.8	38.9	33.9	100.0
1994*	13.6	86.4	–	–	–	–	–
1999*	41.1	58.9	16.0	16.9	35.2	31.9	100.0
2007*	17.5	82.5	7.2	15.1	48.0	29.7	100.0
Respondents' age/ 2018							
18–34	10.7	89.3	11.3	14.4	45.0	29.3	100.0
35–54	12.6	87.4	10.0	16.9	38.9	34.2	100.0
55 and up	13.3	86.7	10.8	17.6	36.7	34.9	100.0
Respondents' education/ 2018							
Elementary	6.5	93.5	11.8	16.8	30.7	40.7	100.0
Vocational	11.8	88.2	11.1	16.5	35.2	37.3	100.0
High school	14.3	85.7	9.5	17.1	43.8	29.6	100.0
Beyond high school	17.5	82.5	8.9	17.8	47.7	25.6	100.0
Farm size/ 2018							
1–5 ha	9.3	90.7	9.3	16.7	39.3	34.7	100.0
6–10 ha	13.5	86.5	11.6	17.9	36.2	34.4	100.0
over 10 ha	22.0	78.0	12.8	15.9	40.9	30.4	100.0

Source: Gorlach, 2009, 2021.

often correlated with higher education. Among those who are the most educated, there is also the highest percentage of respondents who note the positive role of farmers' and agricultural organisations in defending their interests. The two tendencies described above indicate that the progression of agricultural modernisation, as well as the increase in farm size/area, steer farmers' activities toward membership in producers' associations and the search for defenders of their own interests in such organisations. This fits the model of institutionalised collaboration.

The above reflection is confirmed by the data on the effectiveness of particular forms of advocacy for farmers' interests on the right side of the table. Here, there is also the possibility to compare these data with earlier periods of time, such as the 1999 study. Several conclusions might be formulated on such a basis. First, according to the surveyed farmers, the most effective forms of fighting for their interests involved organising in producer groups and cooperatives. Disregarding the "Does not make sense to do anything" answers, the largest category of respondents in all study editions starting in 1999 consists of farmers acknowledging the greatest effectiveness of producer groups and organisations. It is worth mentioning that the highest share of respondents with such opinions was noted in 2007 (48%), and in 2018 they made up 38.9% of all respondents. The significant drop in support for farmers fighting for their own interests through membership in producer groups and cooperatives, noted from 2007 to 2018, when the later study edition was conducted, is not to be ignored. It appears that this might be the effect of Poland's EU membership and the use of its financial instruments encouraging membership in organised groups and associations as well as cooperation between them, which was reflected in the 2007 study results. The novelty of this kind of collaboration wore off and became less meaningful a decade later, and this was confirmed by the data from 2018. This, however, might be considered a general tendency, as the prevalence of such opinions is still valid and acquires special meaning in the context of the evaluation of direct actions. In 2018 the acceptance of direct actions was at the lowest level in comparison to other possible choices. A similar trend could be observed in earlier editions of the study. A low point in support for such activities came after 1999, when it was still the highest in the entire studied period, and before 2007. This could be explained by the special character of the year 1999, when farmers engaged in an unusually high number of contesting activities and had many of their demands satisfied. Thus, the respondents evaluated the effectiveness of these kinds of activities positively at that time.

The general tendency presented here indicates the dominance of positive evaluations of membership in producer organisations and cooperatives as an example of development of the interest representation model based on organised collaboration. This is enhanced by more detailed analysis, taking into consideration additional variables. The tendency for increased support for these types of activities among respondents with a higher level of education and those owning farms of a greater area is also visible. Furthermore, acceptance of protests and the age of respondents are inversely proportional, with younger respondents more inclined to support traditional types of protests.

To conclude, it can be stated that those farmers' activities taken in defence of their own interests, and perceived by them in institutionalised terms, such as producer associations and cooperatives, appeared to be most effective. This should be seen as support for organised collaboration and, to some extent, as its manifestation. The previously described tendencies of farmers' protests generally mirror those of other contemporary social protests. Also, the consequences of farmers' protests, as well other types of protests, lead to the inclusion of protesters' demands in the realm of institutionalised policy. Therefore, it can be stated that, in the 21st century, the model of interest representation in Polish agriculture as organised collaboration is in demand.

Discussion about the model of institutionalised collaboration in Polish agriculture

The evolution of farmers' protests and the prevalence of contesting activities occurring in the area of institutionalised policy are among the most important tendencies confirming the development of the institutionalised collaboration model in relations between the state and farmers and agricultural producers in Poland. The changes observed in protests show that this kind of activity, which is placed outside of institutional policy, loses its importance but does not disappear completely. Protests of this socio-professional category are still present in the public realm, but their status changes significantly. While in the 1990s they were the main form of farmers' influence on decision-making processes of the state, they should now be recognised as just an auxiliary tool, reserved, so to speak, for special occasions. Four facets of farmers' protests could be identified to get to such a point.

First, as previously mentioned, farmers' protests went through the processes of professionalisation, routinisation and institutionalisation. This was bound to cause a decline in their effectiveness and, at the same time, make them more civilised, eliminating many spontaneous and volatile actions. Their force and intensity (range of protest actions, number of participants, and average time of protest) declined in the following years of the 21st century, and their former propagators recognised them as insufficient for fulfilment of their interests.

Second, the diversification of farmers' interests also weakened the effectiveness of their actions. The farmers' financial situation was also worsening as some agricultural sectors were facing crises, thereby impacting farmers' incomes. At different times, various categories of agricultural producers were affected by income drops. As a result, farmers' protests of the 21st century did not come in characteristic waves and, even if some waves were detected, their curves would be flattened compared to the protest waves of the 1990s. Protest activities were still in the picture in subsequent years, but they were not as intense or prominent as the protest waves described earlier. The modernisation of Polish agriculture was an important factor, turning some farmers into agricultural producers or entrepreneurs. For such categories of farmers, protests did not appear to be attractive forms of advocacy for their interests.

Third, some legal changes in the first decade of the 21st century were crucial to the development of the model of interest representation based on organised collaboration. They accelerated and later petrified this kind of relationship between the state and agricultural producers. The first important matter was the introduction of the law on producer groups and associations in 2000. Next was the 2004 law on the regulation of some agricultural markets, which strengthened the position of agricultural producers in the food chain that could do without middle-men selling food. These factors reduced the propensity for protesting due to farmers acquiring a stronger market position and the ability to solve many challenging issues through negotiations.

Fourth, when the Samoobrona union, one of the main organisers of farmers' protests in the 1990s, became a political party and gained parliamentary representation in 2001, the dynamics of interest representation changed further. On one hand, this allowed problems of agriculture to be moved to a greater extent to the forum of institutionalised politics; on the other, it could be seen as a manifestation of the system of organised collaboration progressively consolidating. Finally, there is a document entitled "The Pact for Rural Areas" (*Pakt dla wsi*), which is worth mentioning. Presented to the government in 1999, it included numerous protester demands, as well as proposals for system-wide solutions. Although it was not implemented at the time, it initiated a good practice for subsequent governments to acknowledge and even embrace such types of documents, which presented comprehensive solutions to the various problems of agriculture and rural areas. This is also an example of the organised collaboration model.

Considering the above arguments, it can be stated that farmers noticed the strength of the described institutional solutions and this affected their perception of protest activities as having little effectiveness. Nevertheless, it should be emphasised that the protests were the engine for the development of the interest representation system, based on organised collaboration. Mostly, they allowed farmers to gain agency and some empowerment in the first years of the transformation. Furthermore, demands to strengthen farmers' market position, requests for guarantees of institutional protection for farmers and preparation of an institutional path for conflict solution were among important postulates of the protesting farmers in the second half of the 1990s. Taking into consideration the results of quantitative studies, partially presented in this text, it might be added that unless there are some unpredictable circumstances, such as a collapse of the democratic system in Poland (which cannot be completely ruled out as a possibility), the system of organised collaboration in agriculture will solidify. Cooperation within the framework of producer associations has met with a significant level of support among the younger, well-educated generation of farmers.

In view of the arguments presented here regarding the role of farmers' protests in shaping the model of interest representation based on organised collaboration, attention should be given to two other matters, often omitted in analyses of protests actions conducted by this social group. The first deals with empowerment of peasants and farmers. It should be remembered that these protests were something more to peasants than just an opportunity to fight for their interests. The protests provided them with a platform to have their voice heard in nationwide social discussions. Such a tradition goes back to the beginning of the 20th century, when peasants fought for equal rights within Polish society and the nation. It was present during the time of real socialism, when peasants wanted to retain their autonomy and status as land owners by defending their right to private ownership of land. After 1989 they were able to permanently place their own interests within the framework of institutionalised policy, thanks to protest activities.

The second matter deals with the protests serving in some sort of substitute role for everyday activities within the framework of civil society and their meaning for the consolidation of democracy. This issue has some tradition in academic literature, and particularly in the reflection on democracy consolidation in Central and Eastern Europe (see Greskovits, 1998; Ekiert & Kubik, 1999). Greskovits states that protests can be threatening for the stability of democracy and therefore a certain "silence of society" is desired, especially if there is an alliance of elites overseeing it. This could result in the development of a corporatist order. In other words, democracy without active civil society and protest activities takes a more corporatist form in Central and Eastern Europe, but is more stable because of that fact. However, Ekiert and Kubik suggest something completely different: they do not treat protests as a threat to democracy; rather, they think that they play a positive role in the consolidation of the political and economic system. Intensification of protest in democratic states in this part of Europe could be treated as an indicator of the potency of civil society, which strengthens the democracy. This is a less conservative view of civil society which, besides activities for the common good, also takes into consideration the fulfilment interests of concrete social groups, as this might benefit society as a whole (see Jacobsson & Karolczuk, 2017).

Conclusion

The role of farmers' protests in building a model of organised collaboration manifested in two ways. First, the internal dynamics of these activities led to professionalisation, routinisation and, consequently, institutionalisation, which, on the one hand, took away their element of volatility and compromised their effectiveness to some extent but, on the other, brought farmers' protests into the fold of institutionalised politics. Second, these protests provided an impulse to prepare legislative solutions, thanks to conventional politics and a frame for the model of organised collaboration. Two phenomena played indirect, but important, roles in these processes. The first one, being the stronger of the two, was connected with the processes of the modernisation of Polish agriculture and the stratification of the social category of farmers, as well as diversification of their interests. This had a negative impact on the strength and intensity of protests in the 21st century. Currently, farmers perceive them to be ineffective actions, barely fulfilling a supplementary role in regards to permanent cooperation with the state. If they occur, they constitute a "last resort" form of pressure and are no longer a permanent element of fighting for farmers' interests.

The second phenomenon could be less meaningful, and might be noticed only in the longer historical perspective of farmers' quest for empowerment. In practice, over the last century, farmers had to take up their fight anew in each political system in Poland. At the beginning of the 20th century, peasants strived to gain recognition and equal rights as a permanent and valuable component of the nation. This was meant not only to ensure political representation and civil rights, but also led to agricultural reform in accordance with their expectations. At the time of real socialism, farmers had to take up an uneven, but victorious, fight to keep the farmland in their hands. At the beginning of the transformation, in the free market system which they pioneered, they were among the socio-professional groups in Poland that had to face the highest economic costs of change. However, thanks to their protest activities, they were able to defend their own interests and create favourable conditions for further development. The system of organised collaboration that forms the basis of relations between the state and protesting farmers is an example of that, and of their aim. As this chapter ends, one question remains open and it should provide a stimulus for other studies: To what extent is this system sustainable and viable?

The problem posed in this chapter can be placed in a broader theoretical context. Two aspects of it seem relevant and timely. The first concerns the growing role of non-institutional politics, which is caused by many factors, one of which is the contemporary crisis of democracy (see Rucht, 2007; Della Porta & Diani, 2009). Protest activities in many societies, carried out by various interest groups and ordinary citizens, show that this policy area is gaining in importance and allow effective influence upon decision-making process. Social movements and protests are a good example of this. McAdam, Tarrow, and Tilly (2001) called this type of activity a "contention politics". This develops the notion of protest and collective contestation, goes beyond the traditional political dispute within an institutionalised policy (*contained contention*) and is an example of *transgressive contention*. It can be cautiously assumed that the system of interest representation described here is conducive to strengthening the role of this policy in developed democratic systems. This is due to the role played by the farmers' protests as a factor in the emergence of this type of collaboration model. As a result, contestation becomes a permanent element of a democratic political system, not threatening it, and even strengthening it if it is within the framework of legal norms.

The second theoretical aspect, concerns the generally understood protest activity located in the context of civil society. The low membership of farmers in Poland to organisations in the form of associations and political parties could testify to the weakness of this society, especially in the countryside. It is, however, apparent, as Jacobsson and Karolczuk (2017) argue. According to them, the perspective of civil society theorists' understanding of this as the activity of individuals in organisations is insufficient to describe civil societies in post-communist countries. It is justified to say that these societies are weak in organisational terms. However, if we take as their strength the scope of achieved political or economic goals, it transpires that protest activities bring similar results to "traditional" civic activity. Not only do they allow the protesters to achieve their goals, they also create cooperation networks and various new forms of organisation. In other words, the protests by farmers, as well as other social categories in Poland, can be treated as a manifestation of civil society understood as a dynamic process and not a static set of organisations. This statement evokes the question of whether such a state of affairs concerns only the societies of Central and Eastern Europe, or whether it is a more universal phenomenon.

Notes

1 In 1988 there were 2,534,000 farms in Poland; in 1996 there were 2,041,600 (a decrease of 19.4%); in 2005: 1,782,300 (a decrease of 12.7% compared to 1996); and, finally, 1,425,000 in 2018 (a decrease of 20.0% compared to 2005). Between 1988 and 2018, the total decline in the number of farm was 43.8% (Annual Central Statistical Office – GUS).

2 Study conducted with the research project of the National Science Centre No. 2015/18/A/HS6/00114, under the supervision of Krzysztof Gorlach.

3 NSZZ RI Solidarność – Independent Self-governing Trade Union of Individual Farmers.

"Solidarity"– trade union of farmers that came into existence as the part of the social movement of the "Solidarity" initiated in August 1980.

Samoobrona – Trade Unions of the Agriculture Self-defence, established in 1992 and later converted into a political party named "Self-Defence of the Republic of Poland". It mostly attracted the owners of big farms.

KZRKiOR – National Association of Farmers and Farming Organisations – post-socialist trade union of farmers.

Bibliography

Della Porta D. and Diani M. (2009). *Social movements. An introduction*. London/ Oxford: Blackwell Publishing.

Ekiert, G. and Kubik, J. (1997). Protesty społeczne w nowych demokracjach: Polska, Słowacja, Węgry i Niemcy Wschodnie (1989–1994). *Studia Socjologiczne*, 4, 21–59.

Ekiert, G. and Kubik, J. (1999). *Rebellious civil society. Popular protest and democratic consolidation in Poland, 1989–1993*. Ann Arbor: University of Michigan Press.

Foryś, G. (2008). *Dynamika sporu. Protesty rolników w III Rzeczpospolitej*. Warsaw: Wydawnictwo Naukowe Scholar.

Gorlach, K. (2009). *W poszukiwaniu równowagi. Polskie rodzinne gospodarstwa rolne w Unii Europejskiej*. Kraków: Wydawnictwo Uniwersytetu Jagiellońskiego.

Gorlach, K. (2021). *Think locally, act globally: Polish farmers in the global era of sustainability and resilience*. Kraków: Wydawnictwo Uniwersytetu Jagiellońskiego.

Greskovits, B. (1998). *The political economy of protest and patience. East European and Latin American transformations compared*. Budapest: CEU Press.

Hausner, J. (1996). System polityczny i integracja europejska Polski. In M. Belka, J. Hausner, M. Marody and M. Zirk-Sadowski (Eds). *Polska transformacja w perspektywie integracji europejskiej* (pp. 180–226). Warsaw: Fundacja im. Friedricha Eberta.

Imig, D. and Tarrow, S. (2001). Mapping the Europeanization of contention: Evidence from a quantitative data analysis. In D. Imig and S. Tarrow (Eds). *Contentious Europeans: Protest and emerging polity* (pp. 27–49). Lanham, Maryland: Rowman & Littlefield, Inc.

Jacobsson, K. and Karolczuk, E. (Eds.) 2017. Introduction: Rethinking Polish civil society. *Civil society revisited. Lessons from Poland* (pp. 1–38). New York/Oxford: Berghahn.

Koopmans, R. and Rucht, D. (1999). Introduction to special issue: Protest event analysis – where to now? *Mobilization*, 4(2) (pp.123–130). San Diego: San Diego State University.

McAdam, D., Tarrow S. and Tilly C. (2001). *Dynamics of contentions*. New York: Cambridge University Press.

Pitkin, H. (1967). *The concept of representation*. Los Angeles: Los Angeles University Press.

Roczniki Statystyczne GUS (n.d.). *Bank Danych Lokalnych (Statistical Yearbook Main Statistical Office, Bank of Local Data)*.

Rucht, D. (2007). The spread of protest politics. In R. J. Dalton and H. D. Klingemann (Eds.). *The Oxford handbook of political behavior* (pp. 708–723). Oxford/New York: Oxford University Press.

Schmitter, P. C. (1979). Still the century of corporatism? In P. C. Schmitter and G. Lehmbruch (Eds). *Trends toward corporatist intermediation* (pp. 7–52). London: SAGE Publications.

Tilly, C. H. (1995). *Popular contention in Great Britain 1758–1834*. Cambridge: Harvard University Press.

Wasilewski, J. (1986). Społeczeństwo polskie, społeczeństwo chłopskie (Polish Society, Peasants Society). *Studia Socjologiczne*, 3 (102), 39–56.

13 Rurality

From the margins to the focus of interest

Pavel Pospěch, Eirik Magnus Fuglestad and Elisabete Figueiredo

Rural sociology, and rural studies more generally, are inescapably located at the intersection of social structure and culture. Ever since Sorokin and Zimmerman's (1929) "compound definition", the idea that life in small settlements or in places with low population density somehow correlates with certain beliefs, values and character has created all kinds of misunderstandings. Many of those are rooted in the position of the rural as the Other to the predominantly urban society. Consider, for instance, the way that rural poverty and deprivation can be reframed as a simple and modest lifestyle (Woodward 1996): here, elements of social structure are mistaken for culture. A certain way of living is acknowledged as a specific cultural artefact, even if there are very specific structural reasons which made this way of living a necessity.

In this book, we have encountered the opposite case more frequently: the mistaking of culture for structure. Certain values, sentiments and beliefs are presented as attached to rurality, as its structural feature – yet are they really? Does a feeling of resentment and a sense of being left behind automatically attach itself to rural areas? Or is it the case that rurality – as an element of social structure – just found itself in the middle of a large cultural shift in western societies? Are rural voters more conservative because they are rural voters? Are they more authentic? And what does "authentic" mean anyway? In recent years, we have witnessed, on many occasions, how cultural conflicts and ideological battles are arduously twisted and turned in order to make them fit into the narrative of the rural–urban divide. Indeed, we all have seen the maps of the results of the 2016 and 2020 presidential elections in the United States or of the UK Brexit referendum – on the maps, the big cities are painted with a different colour than the rest of the country and the story of the "rural–urban divide" appears victorious: at last, structure and culture have joined forces.

But have they? Why is it the case that rural areas differ from cities when it comes to certain kinds of values, yet not others? What social and political conflicts have the power to trigger this rural distinctiveness? And speaking of distinctiveness, is it really the rural which is the outlier (as the Othering perspective would suggest), or are these big cities, with high levels of tertiary education, cosmopolitanism and service-based economy, responsible for the different colouring of the map? These kinds of questions have been central to

DOI: 10.4324/9781003091714-15

this volume, and the authors have considered them in various national contexts. Whether the contributions are specifically focused on national politics and contexts or not, the central questions about the interplay between structure and culture still remain: how did politics enter the study of rurality? And how did the rural find itself in the middle of politics? In the following we offer some thoughts on this, informed by the chapters in this volume.

The power of the market: consuming the countryside

To answer these questions, we focus on the ways that structural features are made meaningful – or "framed" (Goffman 1974) – culturally. The concept of framing allows the study of cultural layers which are attached to the underlying structural factors – whether the latter include economics, infrastructure, inequality or the simple fact that something takes place in areas considered rural. While culturalist approaches are frequently criticised for their alleged inability to deal with the problems of power (Gans 2007), it is precisely power which interests us here. To proclaim a certain state of things as natural, or "authentically rural", is an act of power and it requires symbolic capital (Bourdieu 1984) in the hands of those who make such claims. In Chapter 4, Plüschke-Altof and Annist traced such development in a dual structure of a periphery in Estonia. While some places, identities and symbols of rurality are marked as authentic and promoted heavily through a massive heritage campaign, other parts of the periphery remain just what they are: a periphery. All the usual suspects are there: post-productivism, heritage industry, tourism as a driver of rural development – but they are only there for some. The inequalities, the authors argue, are rooted in neoliberal policies which induce a zero-sum competition among Estonian regions. Looking at this case, we can see power operating through a set of unspoken neoliberal assumptions: entrepreneurship and competition are good. People should take responsibility for their fate.

A different situation of power operating from afar was described in Chapter 10, where Bjarnason, Thorarinsdottir and Gkartzios discussed the portrayals of rurality in Icelandic cinema. They invited us into a dark world of "toxic masculinity, alcoholism, violence, disintegrated communities, harsh nature and broken people unwilling or unable to leave for the city". These images appear strange at first: next to the country's booming tourist industry, the dark scenes of rural life in Iceland don't really fit. Is this some new sort of Nordic self-deprecation? Not really, the authors argue: the dystopic images are what sells Icelandic films to the audiences at international film festivals. There is a demand for stories of despair and isolation amid an imposing Icelandic landscape, and there is a supply, too. Film connoisseurs are looking for these images to provide a sense of the "real Iceland" in the world of independent films, itself opposed to the happy and colourful productions which come out of Hollywood. It would appear that the romantic idealisation of the rural Seto people in Estonia (Chapter 4) and the disparaging images of rural Icelanders are somehow opposed to each other – but they aren't. In both cases, a particular representation of rurality fills a niche in the global market. It is a rurality laid out for consumers: here, watch the Setos perform their dances; there, watch the Westfjords sheep-grazers drinking their way out of their loneliness. In both cases, rural authenticity, both idyllic and anti-idyllic, is being produced, packaged and sold to a crowd of global consumers.

One of the perils of such performances of authenticity lies in their exclusiveness: there is only one "authentic" way of rural life. Rural authenticity links the quantitative, qualitative and performative elements of authenticity (Varga 2011): it strives to establish a local uniqueness which makes us different from others, but the same among ourselves. The truth of the authentic performance does not allow for dissent. Questioning the authentic means asking about the truth values of something that has already established itself as unquestionable. As tourists, this is why we visit the Seto region and marvel at its traditions: they contain an element of unquestioned truth, which is something that, in our urban civilisation, we experience as lost. In Chapter 9, Dünckmann identified this post-political idealisation in the images of rural idyll. Referring to Arendt's (1958) work, Dünckmann sees the kitsch of the rural idyll as a potential agent of the post-political condition, in which the promise of a universal agreement on goodness "beyond dispute" threatens the very core of the democratic process (Mouffe 2005; Swyngedouw 2009). Like in Sennett's (1977) intimate society, kitsch transforms our sensory pleasures and our psychological fears into ethical imperatives. "This is how the world should be and this is how we should live", rural idylls make us think. Indulging in the many market-supplied versions of rural authenticity, we cannot help but experience our own nostalgic desire for wholeness, for truth and for a world where things work beyond dispute.

In a moving commentary in the *Tristes Tropiques*, Lévi-Strauss suggests that the imperfections and residues in the stream of civilisation are what we love and what we cherish the most – yet, at the same time, they are what we are inevitably bound to eliminate. "Social life consists in destroying that which gives it its savour", he concludes (1992: 384). The consumer's desire for the experience of authentic rurality seems to be caught in the same dilemma: we are attracted by the exotic, knowing that, through our own way of life, we are gradually moving towards its elimination. This contrast, which informs many modernist rural policies, has been labelled as a politics of defence (Bell 2007). In Chapter 7 of this volume, Flø makes a powerful statement against this. His declaration attacks the very idea of rural areas as places of consumption. The turn towards selling rurality as a product for the middle classes, which, in a number of developmental programmes – such as the Estonian one – features as a solution for the lagging rural areas, is what Flø identifies as the problem.

The men in ties: the rural and the urban

Flø's chapter is a meeting point of two large, underlying themes of this book: the consumption of rurality and the rural–urban divide. Flø argues that today's Norwegian villagers are bitter because they feel they had been cheated once already: the tourism- and consumption-led development was a scam. They have shut industries, taken away the jobs – and now they're coming back for more as centralisation continues. While the anger is easy to understand, it is less easy to see who exactly are "they". The urbanites? The globalisers? The capitalists? Fair enough, but isn't the whole world in which we live urbanised, globalised and capitalist-controlled?

This is one of the key features of rural distinctiveness: in political conflicts, rurality presents an almost universal antithesis. In one way or another, all of these conflicts – the rural–urban divide, anti-globalism, etc. – are inscribed in the underlying tension between the centre and the periphery (Shils 1975), but their specific instances may differ. Thus, one can refer to the values of the periphery when arguing for the protection of the environment – but one can refer to the same values when arguing for its destruction, like rural Americans do in Hochschild's *Strangers in Their Own Land* (Hochschild 2018). Nikulin and Trotsuk's (Chapter 6) brief review of the various forms of "populism from above" in Russian regions is a case in point: the voices from the Russian periphery provide a colourful mix of anti-centre sentiments. The feeling of peripherality, of being left behind, is what unites these voices. Hearn's contribution (Chapter 2) to this volume opens the discussion of being left behind from a historical perspective. The process of nation-building, Hearn argues, fosters a never-ending negotiation about who *the people* are. Throughout this process, rurality has been pushed further into the periphery: from a somewhat distant, yet recognisable alternative to the urbanism of the mid-20th century, rural life has moved into the utopic and the unreal. The tension between the urban and the rural has become anachronic, and in the nation-building process, it was replaced with conflicts around multiculturalism and identity politics. At first, Hearn's suggestion appears contradictory to the perceived revival of the rural–urban divide as it is being observed on many occasions throughout this book. Yet, the contradiction is not so simple: the "old" rural–urban conflict that Hearn refers to was a tension between two ways of life; in popular culture, such tension might have been expressed through simultaneous "condescension and sentimental admiration" of rurality in the American TV shows of the 1960s. The revival, or the "new" rural–urban conflict, is played in a different key: what kind of rurality is invoked when Trump, Orbán and Kaczynski praise the villagers and bash the urban elites? Is this still a rurality and still a way of life? Or are these reactionary values which are being brought up under the guise of rurality? With this question, we are back at the structure–culture junction: does the "rural" in the current rural–urban divide have anything do with rurality as structural feature, or is it, rather, a political representation, a cultural artefact, deployed in the battle for a cultural backlash (Norris and Inglehart 2019)?

Whatever the answer is, the rural–urban divide has become a highly visible motif in the public discourse, restated with each major electoral event. As the populist vote, which mostly appears to come from non-urban and peripheral areas, is often framed as "protest vote" (Van der Brug et al. 2000; Cramer

2016), the very idea of *rural protest* comes back into focus. The emergence of the *Gilets Jaunes* (yellow vests) movement in France in 2018 was a striking form of the revolt of the "nameless" periphery against the urban centres. The movement originated with a protest against rising fuel prices and stricter speed limits for drivers – both directly affecting those whose livelihoods are dependent on driving a car, in a country where not needing a car has become a status symbol for the wealthy urban class.[1] The grievances, and their symbol – the yellow vest – carry a strong message of authenticity. Rather than being an artificial symbol, the vest accompanies the workers in their everyday lives, often highlighting the fact that they work under difficult and dangerous conditions where a high-visibility vest is required for protection. The movement is organised in a strictly horizontal way and its "anti-politician" ethos prevents leaders from emerging from within (Grossman 2019), emphasising the movement's ambition to speak for *the people*.

In Woods' study (Chapter 3) of rural protest in the UK, a similar process of frame extension (Snow et al. 1986) to that of the yellow vests can be observed. In France, the protests were sparked by a seemingly minor fuel price change and a speed limit adjustment. In the UK, the same role was played by hunting: a ban on hunting wild animals with hounds started a massive wave of rural protest in the late 1990s and early 2000s. For a frame extension to take place successfully, cultural work must be done: driving – or hunting – must be presented as a central component of the rural way of life. A fight for a freedom of driving and hunting thus becomes a fight for authentic rurality and, more importantly, it becomes a fight against those who are seen as oppressing the rural way of life. Woods' study argues that there is no clear continuity between the rural protests at the turn of the millennium and the 2016 Brexit referendum, which brought the UK out of the European Union on the back of a strong anti-elitist sentiment. However, the protests have entrenched the "rhetoric of an urban polity out of touch with rural people", a trope which became prominent in the politics of the rural–urban divide. The Brexit campaign then came to harvest the fruits of this long-established conflict.

Authenticity is a central feature of rural protest. The stories included in this volume – whether of the yellow vests, the betrayed farmers in Norway, or the defenders of hunting in the UK – are always stories of people positioned against the inauthentic, out-of-touch elite, shielded from real life in the urban centres. The inauthenticity of the armchair experts can posit a stigma which disqualifies them from being accepted into the public debate. In this sense, Perrotta's study of the Italian debate on the *regularisation* of migrant workers (Chapter 5) is highly illuminating. Perrotta traces the political discourse of the three leading actors of the debate: the centre-left Minister of Agriculture, the populist radical and the grassroots union organiser. Despite strong political differences, the three are united by highlighting their personal experience and knowledge of life in farms. This *populism as a method* makes the actors constantly remind their audiences of their authentic ties to the rural periphery, ending in the populist and the union organiser one-upping each other about who has personally visited more farms. This insistence on presenting oneself as "one of the people" stands in some contrast to Foryś's (Chapter 12) observations from Poland, where rural protest has taken some steps away from the farm. After waves of unorganised, often spontaneous farmer protests in the post-socialist transformation era, the protests have become more organised and institutionalised in formal structures of representation. The farmers' identification with the Polish peasant tradition has waned over this period, with traditional symbols of peasantry such as scythes and sickles disappearing from the rallies – while, at the same time, the nationalist and religious symbols have been retained by the protesting farmers. Taking Perrotta's observations into account, one might ask whether the professionalisation of Polish rural protest does not pave the way for alienation; and if, consequently, an emergence of a more "authentic" wing of protesters is not to be expected.

The fact that authenticity is a political currency becomes very visible in Chapter 8, where Bosma and Peeren discuss the #Proudofthefarmer movement in the Netherlands. In a key part of the analysis, Bosma and Peeren point out the framing of farming as identity: farming is not something you *do*; it's something *you are*. The farmer is authentic not just because he (it's rarely a *she* in the movement) lives close to the land and nature. He is authentic because his job is his calling, his Weberian *Beruf*. You are a farmer at heart, even if you find yourself working in a different business eventually. The sincerity of such identity can hardly be questioned. On the opposite side of the trench, there are "government politicians and bureaucrats" whose "jobs are external to their being, like the ties they wear". *The Men in Ties*: the Dutch farmers could not dream up a better set of antagonists for their cause. They can change their ties every day, like they change their opinions. Their ties won't get dirty, because the work they do is not real enough. They always have one foot out of the door, as they are catching a ride or a flight to one of their meeting – while the farmers remain in one place, tied to what is truly theirs, and what can't be changed as simply as the colour of one's tie. Like Arendt and Dünckmann, Bosma and Peeren point out the questionable consequences of elevating the "good things" (Ahmed 2008), like the authenticity of farming life, to political principles.

To borrow some concepts from Alexander's (2006) Civil Sphere theory, the authenticity of rural life is anchored in both the deep structures of the civil discourse and its temporal structures. For the former, the representations of rurality are attached to the civil pole of the discourse as true, honest and self-sufficient. On the opposite side of the discourse, the men in ties are portrayed as deceitful, calculating and invasive. For the latter, the narrative of rural authenticity is strongly rooted in history – but there are changes that must be accommodated, too. In Chapter 11, these are traced by Kristensen and Gullikstad in their study of immigrant integration in Norwegian villages. Drawing on the concept of place-making narratives (Aptekar 2017; Junnilainen 2020), the authors show, once again, how rurality is reframed into new conceptual constellations including heterogeneity and cosmopolitanism.

There is an underlying message of success here: rural areas have the ability to do better than cities when it comes to the integration of immigrants. This may seem at odds with the conservative narrative of closed rural communities which cannot accommodate change because, like the Dutch farmers, the people "are what they are". On the other hand, it does show the strength of the community. Among the Europe-wide fearmongering which followed the migration wave of 2015, these Norwegian rural areas appear to be just what our fantasies have projected them to be: places of stability and functioning, authentic community.

***

At the time of writing and submitting this book, borders around the world are being re-drawn by the COVID-19 pandemic. This includes the borders between the urban and the rural. The disadvantages that rural populations have been experiencing with regards to infrastructure and services are magnified by the pandemic. Insufficient access to hospitals, health care and testing infrastructure, together with reliance on transport networks, push those who have been left behind even further behind. In many countries, rural populations are generally older than their urban counterparts, and the concentration of risk groups is higher in rural areas, too. The rural–urban divide takes on a new shape as governments come up with restrictive measures regarding work, mobility and social life: in many European countries, middle-class urbanites were seen leaving cities for their second homes in rural areas. As the spread of the disease has reached the highest numbers in urban centres thus far, these urban refugees faced a cold welcome by the rural inhabitants, who had a good reason to see them as potential carriers of the virus.

To end on a positive note, this middle-class urban exodus carries a different message, too. As services, shops and social activities are locked down in many countries, life in cities turns its ugly face towards us: the face of social isolation, of small living spaces and of the looming dangers of a still vastly unknown disease. The economic and social blow might be the same for rural areas, but in terms of everyday experience the rural populations appear to lose less than their urban counterparts. After all, what is life in a city good for, when the shops, cafés and museums are all closed, and when meeting friends becomes a health hazard? While the market for urban housing is stagnating, the demand for out-of-town houses has increased rapidly. The pandemic forces us into the online world, yet it also makes us appreciate the tangible: nature, free space, open sky. Somehow, this is reminiscent of the Dutch case, with farmers standing against the "men in ties": the men in ties *do what they do*, the farmers *are what they are*. Similarly, to be in a city means to *do things*: to meet others, to visit places, to sit in cafés and restaurants. Without *doing* city life, *being* in a city feels empty. In comparison, in rural areas, perhaps one doesn't need to *do* so much. Perhaps *being* is just fine. Or, at least, this may be one message that the global pandemic brought us. We would prefer to learn it in a different way, but that is not open to us to choose.

Note

1 www.nytimes.com/2018/11/24/world/europe/france-yellow-vest-protest.html

References

Ahmed, S., 2008. The politics of good feeling. *ACRAWSA e-Journal*, 4(1), pp. 1–18.

Alexander, J. C., 2006. *The civil sphere*. Oxford: Oxford University Press.

Aptekar, S., 2017. Looking forward, looking back: Collective memory and neighborhood identity in two urban parks. *Symbolic Interaction*, 40(1), pp. 101–121.

Arendt, H., 1958. *The human condition*. Chicago: University of Chicago Press.

Bell, M. M., 2007. The two-ness of rural life and the ends of rural scholarship. *Journal of Rural Studies*, 23(4), pp. 402–415.

Bourdieu, P., 1984. *Distinction: A social critique of the judgement of taste*. Cambridge: Harvard University Press.

Cramer, K. J., 2016. *The politics of resentment: Rural consciousness in Wisconsin and the rise of Scott Walker*. Chicago: University of Chicago Press.

Gans, H. J., 2007. But culturalism cannot explain power: A reply to borer. *City & Community*, 6(2), pp. 159–160.

Goffman, E., 1974. *Frame analysis: An essay on the organization of experience*. Cambridge: Harvard University Press.

Grossman, E., 2019. France's yellow vests – symptom of a chronic disease. *Political Insight*, 10(1), pp. 30–34.

Hochschild, A. R., 2018. *Strangers in their own land: Anger and mourning on the American right*. New York: The New Press.

Junnilainen, L., 2020. Place narratives and the experience of class: Comparing collective destigmatization strategies in two social housing neighborhoods. *Social Inclusion*, 8(1), pp. 44–54.

Lévi-Strauss, C., 1992. *Tristes tropiques (1955)*. London: Penguin Books.

Mouffe, C., 2005. *On the political*. New York: Psychology Press.

Norris, P. and Inglehart, R., 2019. *Cultural backlash: Trump, Brexit, and authoritarian populism*. Cambridge: Cambridge University Press.

Sennett, R., 1977. *The fall of public man*. New York: A. Knopf.

Shils, E., 1975. *Center and periphery*. Chicago: University of Chicago Press.

Snow, D. A., Rochford Jr, E. B., Worden, S. K. and Benford, R. D., 1986. Frame alignment processes, micromobilization, and movement participation. *American Sociological Review*, 51(4), pp. 464–481.

Sorokin, P. A. and Zimmerman, C. C., 1929. *Principles of rural–urban sociology*. New York: Henry Holt and Company.

Swyngedouw, E., 2009. The zero-ground of politics: Musings on the post-political city. *New Geographies*, 1(1), pp. 52–61.

Van der Brug, W., Fennema, M. and Tillie, J., 2000. Anti-immigrant parties in Europe: Ideological or protest vote? *European Journal of Political Research*, 37(1), pp. 77–102.

Varga, S., 2011. The paradox of authenticity. *Telos-New York*, 156, pp. 113–131.

Woodward, R., 1996. "Deprivation" and "the rural": An investigation into contradictory discourses. *Journal of Rural Studies*, 12(1), pp. 55–67.

Index

Note: Page numbers followed by n indicate notes.

Adorno, T. 114–115, 117–118, 132
agrarian romanticism 135
agricultural exceptionalism 29
agricultural reform 30, 184
Ahmed, S. 115, 121–122
Albatross (film) 152n4
Alexander, J. C. 192
Almås, R. 96–97
Andersen, D. T. 105
Anderson, B. 20
Angry White Men (Kimmel) 120
Annist, A. 8, 42–55, 188
anti-globalisation protests 28
anti-government protests 31
anti-hunting legislation 30
anti-idyll 130–131
anti-immigration politics 115, 119, 121
anti-Putin protest 81
Aptekar, S. 160
Arendt, H. 9, 129, 133–135, 137, 189, 192
asylum-seekers 157–158
authentic farmer 114
authentic rurality 2, 6–7, 42, 53, 189, 191

Bakhtin, M. 117–118
Baldwin, S. 35
Balling, E. 144
battle for refugees 156, 158
Baudet, T. 118–119, 121
Baumann, C. 43, 130
Being and Time (Heidegger) 117
Bellanova, T. 60–62, 65–67, 69–72, 74n5
Berlusconi, S. 64
Bernhard, T. 131
Between Mountain and Shore
    (Gudmundsson) 144
Binkley, S. 132

Bjarnason, T. 9, 141–152, 188
Black's Game (film) 152n6
Blood-red Sunset (film) 152n3
blue collar 21, 106
Boomsma, D. 122
Bosma, A. 9, 113–125, 192
Bourdieu, P. 53
Brahmin left 5
Brave Men's Blood (film) 152n6
Brexit 2–3, 7–8, 15–16, 21, 23, 27–39,
    187, 191
Bristow, G. 49
British National Party 32
Broch, H. 132
Brooks, S. 28, 39
Brubaker, R. 3, 77
Brundtland Commission 101
bureaucratic leadership style 81

Capital and Ideology (Piketty) 5
Capital in the Twenty first Century
    (Piketty) 5
Castberg, J. 98
Central-Eastern European (CEE):
    market-economy reforms in 46;
    peripheralisation of 42
centre–periphery relations 4–6
Chesterton, G. K. 37
Children of Nature (film) 148, 152n5
Chopra-Gant, M. 130
city and countryside: classic nation
    building 15, 17–19, 22, 24; nation
    deconstructing 16, 19–22
City State (film) 152n6
civil rights movements 20, 184
civil society 15, 24, 63, 84, 89n2, 100,
    133, 176, 183–184

class conflict 24
classic nation building 15, 17–19, 22, 24
Clinton, H. 15–16
Cloke, P. 97
commodification and populism 1–10
Common Agricultural Policy (CAP) 32
communication 16, 18, 20, 22
Communist Party 79, 81
community police reform 107
Conservative Party 28, 32–33, 38
consumerist idyll 43
contemporary populism 78
contemporary societies 7, 87
contention politics 184
contract farm 30
core–periphery 27
cosmopolitanism 156–157, 162, 166, 187, 192
counter-hegemonic imagery 62–64
counter-urbanisation 2, 43, 157
Countryside Alliance 28–29, 31–34, 35, 36–37, 38
countryside protest movement 28–29, 31–39
Country Wedding (film) 145
COVID-19, 7–8, 193; Italy, health crisis 60–73; lockdown 115, 123–124
Cramer, K. 3
Cucco, I. 63
cultural elite 45, 53–54, 77, 94
cultural policy, Norway 96–97

The Daily Telegraph (newpaper) 36
De Gelderlander (newspaper) 114
de Groot, T. 113–114, 118
demographic refill 157
De Volkskrant (newspaper) 120
disempowerment of farmers 119
Dis (film) 145
dispossession 42–43, 52; accumulation by 45, 52–53; neoliberalisation of rural authenticity 51–53; of rural population 43
Dryzek, J. 102
Dünckmann, F. 9, 129–138, 189, 192
Durkheim, E. 18
Dutch farmers' protests: anger of 120–123; authenticity and nationalism 115–118; protesting farmers as populist heroes 118–119; rural, politics of 124

East (film) 152n6
Eckersley, R. 102

ecological modernisation 94, 101–102
economic dispossession 52–53
economic elite 48, 62, 72
economic equality 5
economic globalisation 1, 16, 19
economic growth 4–5, 17–18, 46–47, 101, 157–158, 166
economic marginalisation 28
economic rationalism 102
economic reforms 46, 177
Either Way (film) 152n4
Ekiert, G. 171, 183
electoral geography, politics of 29–31
environmental economics 102–103
Eriksson, M. 150
Erlingson, B. 148
Estonian Conservative People's Party (EKRE) 54
EU membership 35–36, 174, 180
Europeanisation 174
European Union (EU) 27, 32–33, 36, 38, 54, 174, 191
Euro protest 174

Farmers Defence Force (FDF) 113
farmers' protest movement 31–32
farmers unions 29, 38, 173, 178
farm labour, Italy 62–64
Ferguson, A. 24
Figueiredo, E. 1–10, 187–194
film industry, Iceland 143–145
Flare (film) 152n6
Flø, B. E. 9, 94–107, 189
Fonte, M. 63
food security 29, 65, 80
Forin, R. 63, 65
Foryś, G. 10, 170–185, 192
Friday the 13th (film) 131
Front National 27–28
Fuglestad, E. M. 1–10, 187–194
Fukuyama, F. 5
Furre, B. 98–99

Gardiner, J. 33
gastronationalism 72
gastronationalist 62, 72
Gellner, E. 4, 16, 77
Gemeinschaft and Gesellschaft 18
German Green Party 137
The Girl Gogo (Balling) 144
Giske, T. 105
Gkartzios, M. 9, 141–152, 188
global civil society 24
Goodhart, D. 3, 107

Grudinin, P. 81–83, 89n5
Guilluy, C. 2
Gullikstad, B. 10, 155–167, 192
Guðmundsson, G. A. 144, 148
Guðmundsson, L. 144

Haandrikman, K. 157
Habeck, R. 137
Haga, A. 105
Hajer, M. 101–102
Hakonarson, G. 148
Hals, F. 119
Hanekes, M. 131
Hart, S. 33
Harvey, D. 44
Hausner, J. 176
Hearn, J. 7–8, 15–24, 190
Heartstone (film) 147
The Heath (film) 152n4
Hechter, M. 142
Hedberg, C. 157
Heidegger, M. 117, 123
heritage culture 42–46, 48–49, 54
Hermansen, T. T. 100
Hiddema, T. 118
Hochschild, A. R. 190
Hoey, K. 33
homogenisation 4, 49
Horkheimer, M. 132
Houellebecq, M. 10n1
Howard, N. 63, 65
The Human Condition (Arendt) 133
humanitarian approach 63–64, 72
The Hunters (film) 150
hypercapitalism 5

Icelandic Film Fund 143–144
Icelandic films: film industry 143–145;
   internal and external orientalism
   149–151; rurality of contemporary
   145; social development in 143–145;
   urban and rural 145–149
ideal of authenticity 116
idyllic politics 129, 135–138; and anti-
   idyll 130–131; defending disunity
   against harmony 133–135; kitsch
   131–133
Ikingut (film) 152n2
illegal migrants 64
immigration
Ingaló (film) 152n5
institutionalised policy 180–182, 184
integration and rurality 156–159; labour
   157–158; methods and analytical tools

159–164; positive 160–162; successful
   integration 162–164; xenophobia and
   opposition to 164–165
In the Arms of the Sea (film) 152n5
I Remember You (film) 152n3
"irregular" migrants 71, 74n5
Italian agri-food systems 62–64
"Italian" cultural identity 64
Italy, health crisis: Covid-19 pandemic
   65–66; farm labour 62–64;
   hegemonic and counter-hegemonic
   representations 62–64; "Made
   in Italy" food 61–62; migrant
   workers 65–71; populism as
   method 71–73; quality turn 62–63;
   undocumented migrants 61, 65–66,
   69–72, 74n6

Jacobsson, K. 184
Jansson, D. R. 150
Jar City (film) 152n6
The Jargon of Authenticity (Adorno)
   114, 117
Jóhannes (film) 145
Junnilainen, L. 160

Kailyard School 19
Kari, D. 148
Karolczuk, E. 184
Kasimis, C. 157
Kemp, Bart 114
Kimmel, M. 115, 120–121
kitsch 9, 131–135, 137, 189
Kjelstadli, K. 104
Kristensen, G. K. 10, 155–167, 192
Kubik, J. 171, 183
Kundera, M. 132
KZRKiOR 172, 178, 185n3

labour contract 61
labour immigration 157–158
Labour Party 38, 98–99, 103, 159
labour union 61, 63, 67, 75n11
Laclau, E. 61
Langhelle, O. 101
Lawson, N. 98–99
Leave campaign 28, 33 34, 37
Lenin State Farm 81, 89n5
Le Pen, M. 2, 6
Lévi-Strauss, C. 189
Liberal-Democratic Party 79, 81
Lindgren, A. 137
local authenticity 10

Macho, A. M. 118–119
Maine, H. 18
Major, J. 35
Martina, M. 63
mass immigration 119
McAdam, D. 184
Melnichenko, V. 81–83
melting pot 19, 157
merchant right 5
Midwinter (film) 152n3
migrant workers: denials of asylum
    applications 65; farm labourers
    61–64; Italy health crisis 65–71; living
    conditions of 8, 72; regularisation
    191; sensationalized brutality 63
militant protest movements 30
Mill, J. S. 116
Mississippi Burning (film) 150
model village 6, 54
modern abolitionism 63
Molina, F. 118–119
moral economy 78, 87–88, 89n2
Mouffe, C. 61
multiculturalism 15, 23, 190

narodnik movement 8–9, 78–79,
    83–87, 89n3
national identity 19–20, 35–36, 47, 62,
    144, 150
nationalism: farming and authenticity
    115–118, 123–124; and industrialism
    4; and populism 1, 3–4, 6, 34–37;
    revival of populist politics 3; right-
    wing nationalism parties 27; rural
    protests 34–37; theories of 16; and
    traditionalism 79
nationalist sentiments 2–3
National Party 29, 31–32
National Socialists 135
nation building 4, 7; classic 15, 17–19,
    22, 24; period of 18, 21, 98; principles
    and orientations 7; urban-centric
    process of 15
nation deconstructing 16, 19–22
Nations and Nationalism (Gellner) 4
nature conservation 30
neoliberalisation: case study 46–48; of
    regional politics 8, 42–44, 46; of rural
    authenticity 51–53
neoliberalism: agricultural markets,
    deregulation of 29; and globalisation
    7, 15
neoliberal order 97–99
neo proprietarianism 5

New Life (film) 147
new regional policies, Norway 103–105
Nikkels, M. 116
Nikulin, A. 8, 77–88, 190
Noah the Albino (film) 148
Nói Albínói (film) 152n5
non-confrontational campaigns 30
Norway: cultural policy 96–97;
    deregulating and new regulating 100;
    ecological modernisation 101–102;
    environmental economics 102–103;
    fight for survival 105–106; neoliberal
    order 97–99; new regional policies
    103–105; rural dissatisfaction 94–96,
    98; rural policy 96–97; smouldering
    95–96; social democratic order 97–99;
    sustainable development 100–101
Norwegianist movement 98–99
Norwegian Planning and Building
    Act 100
nostalgic sentimentalism 107
No Trace (film) 152n6
NSZZ RI Solidarność 172, 178, 185n3

Of Horses and Men (film) 147
On Top (film) 147
The Outlaw and his Wife (Sjostrom) 143
Outlaw (film) 152n1

The Paradox of Authenticity 116
Paris of the North (film) 147, 152n4
Parker, A. 150
Peeren, E. 9, 113–125, 192
People of Dugghola (film) 152n2
peripheralisation 42–43, 45, 47
periurbanisation 130
permit of stay 60–61, 66, 68–70, 74n5,
    74n6
Perrotta, D. 8, 60–73, 191–192
Petrini, C. 62
Piketty, R. 5
Pitkin, H. 176
Plüschke-Altof, B. 8, 42–55, 188
Poland farmers' protests: in agriculture
    181–183; evolution 1989–2018,
    171–176; and farmers' consciousness
    178–180; four models 176–178; model
    of institutionalised collaboration
    178–183; protest event 171–172
policy communities 29
Polite People (film) 147
political correctness 20, 34
political entrepreneurship 6

political marginalisation 28
political movement 27, 36, 78, 87
political socialisation 30–31, 37–38
The Politics of Resentment: Rural Consciousness in Wisconsin and the Rise of Scott Walker (Cramer) 3
"pollutor pays" principle 102
populism: "from above" 80–83; "from below" 83–86; institutionalisation of 48–54; as method 71–73; methodological approach and database 45–46; and nationalism 1, 3–4, 6, 34–37; and neoliberal use of rural authenticity 43–45; radical neoliberalisation 46–48; rural protests 34–37
populist anti-immigration 28
populist leaders 80–83
populist movements 8, 37, 42–43, 55, 83, 88
populist sentiments 2–3
positive immigration 160–162
Pospěch, P. 1–10, 187–194
post-socialist neoliberalisation 42, 54
Prince of Darkness (film) 152n1
Progress Party's (FrP) 100
protest campaign 172
#proudofthefarmer 9, 113–125, 192
Putinism 88
Putin's populism 78
Putin, V. 78–83, 87–88

qualitative authenticity 116
Quiet Storm (film) 152n5

racism 20, 23, 130, 150, 156
radicalisation, rural protests 37–38
Rams (film) 148
Rauch, N. 131
red belt 79
"red–green" coalition 103
refugee crisis 155, 159
regional leaders 80–83
Renzi, M. 61
Republican party 28
Rettedal, Arne 99, 101
Reykjavík (film) 147
Reykjavík Rotterdam (film) 152n6
The Road to Somewhere: The Populist Revolt and the Future of Politics (Goodhart) 3
Røed, M. 158
Rovira-Kaltwasser, C. R. 61

Runarsson, R. 148
rural archetype 118–119
rural belt 79
rural conflict 96
rural culture 35, 85, 152
rural discontent 28, 30–33, 35, 37–39, 82
rural dissatisfaction 94–96, 98
rural geography 6–7; politics of 29–31
rural globalisation 156
rural identity 35, 156, 166
rural idyll 7, 9, 30, 43–44, 105, 117–118, 124, 129–131, 136, 147–148, 150–151, 156–157, 189
rural infrastructure 30
rurality 34–37; power of market 188–189; rural and urban 189–193
'rural militia' movement 28
rural policy, Norway 96–97
rural protests 2, 29–30, 191–192; discontent 28, 30–33, 35, 37–39, 82; frame continuity 34–37; nationalism 34–37; populism in rural Britain 31–34; radicalisation 37–38; rurality 34–37; trust 37–38
rural resilience 156
rural resistance 94, 107
rural sociology 1, 6–7, 97, 187
rural–urban divide 1, 6–7, 15, 17–18, 21, 23, 27, 35–36, 94, 124, 158, 187, 189–191, 193
rural–urban relationship 17–19, 21–23, 34, 35
Russian populism 78, 83, 87
Rye, J. F. 156

Sætermo, T. 158
Salvini, M. 61–62, 64
Savchenko, E. 81–83, 89n4, 89n6
Schmitter, P. C. 176
Schouten, C. 113
Scruton, R. 119
Segers, G. J. 113
Sennett, R. 189
Serotonin (Houellebecq) 10n1
service sector 18–19, 21–22, 177
Seto authenticity: cultural heritage elite 53; dispossession 51–53; neoliberal calls 49–51; populism of 54
Setoification 51
settlement of refugees 157–158, 163
Sjostrom, V. 143
small deeds theory 83–86
Smiles, S. 116
Smith, A. 17

social change 24, 30, 150, 170
social conflicts 7, 177
social contract 4
social democratic order 97–99
social dispossession 45
social diversity 156
social exclusion 150, 157
social mobility 4–5, 16, 21
social orders 4, 20
social reform movements 130
socio-political research 45, 94
Søholt, S. 158
Solberg, P. 104
Solidarity 185n3
Sorokin, P. A. 187
Soumahoro, A. 61–62, 71–72
Sparrows (film) 148, 152n5
Spooks and Spirits (film) 147
Steen, A. 158
Strangers in Their Own Land
  (Hochschild) 190
Sundvall, K. 150
supreme arbiter 82
sustainable development 100–101
symbolic dispossession 42, 44–45, 52

Tarrow, S. 174, 184
Tartu Peace Treaty 48
Teigen, H. 97, 103, 107n2
Texas Chainsaw Massacre (film) 131
Thorarinsdottir, B. 9, 141–152, 188
Thought and Change (Gellner) 4
Tilly, C. 184
Tönnies, F. 18, 134
trade unions 171, 173
Treurniet, J. 115
Tristes Tropiques (Lévi-Strauss) 189
Trotsuk, I. 8, 77–88, 190
Trouw (newspaper) 116, 124n2
Trump, D. 2–3, 6, 15–16, 20–21, 23, 27, 190
trust, rural protests 37–38
*Twilight of the Elites: Prosperity, the
  Periphery and the Future of France*
  (Guilluy) 2

UK Independence Party (UKIP) 28,
  32–33, 38–39
The Unbearable Lightness of Being
  (Kundera) 132, 138
Under the Tree (film) 147
undocumented migrants 61, 65–66,
  69–72, 74n6
urban gardening 130
urbanisation 2, 7, 23–24, 43, 47, 86, 130,
  135, 145, 150–151
urban–rural distinction 27
urban–rural divide 43
utilitarian approach 66, 69

Valley Life (film) 147
van Gogh, V. 119
van Keimpema, S. 119–120
van Mulligen, P. H. 115
Varga, S. 116, 121

The Waltons (television show) 130
Wealth of Nations (Smith) 17
Weber, E. 16
white collar 21
The White Ribbon (film) 131
white rurality 156–157
Wilders, G. 118–119, 122–123
Williams, B. 33
Williams, R. 130
wokeness 20
Wolf (film) 152n6
women's rights 23
Woods, M. 8, 27–39, 191
workforce shortage 8, 61, 65–66
working class 19, 21–22

xenophobia 157, 164–166

Zimmerman, C. C. 187
Žižek, S. 135
Zubarevich, N. V. 79
ZZ Samoobrona 172, 178, 182, 185n3

Printed in the United States
by Baker & Taylor Publisher Services